COMMENTARY ON THE ASHI STANDARD OF PRACTICE FOR HOME INSPECTIONS

COMMENTARY ON THE ASHI STANDARD OF PRACTICE FOR HOME INSPECTIONS

Bruce A. Barker

Commentary on the ASHI Standard of Practice for Home Inspections

Publisher's Cataloging-in-Publication Data

Bruce A. Barker

Commentary on the ASHI standard of practice for home inspection / Bruce A. Barker

Includes bibliographic references and index.

ISBN 978-0-9848160-4-0

1. Housing- Standards --Popular works. 2. Building laws--Popular works. 3. Dwellings--Design and construction--Standards--Popular works. I Title.

rev 070518

Commentary Objective and Scope

The primary objective of this commentary is to provide guidance to home inspectors who use the American Society of Home Inspectors (ASHI) Standard of Practice for Home Inspections, version effective 1 March 2014 (HISOP). This commentary is intended to be used by home inspectors. This commentary is not intended to be used by others to interpret or explain the HISOP, to interpret or explain other Standards of Practice (SOPs), or to reduce or expand the scope of the HISOP. In the event of a discrepancy or conflict between this commentary and the HISOP, the HISOP shall govern.

Many home inspection SOPs are based on a version of the HISOP. When other SOPs are similar to the HISOP, home inspectors may wish to use this commentary. It is important to note, however, that when a SOP is mandated by state law or regulation, the state regulatory authority is the best source of guidance about how to interpert and apply the SOP in that state.

The opinions in this commentary are those of the author. They are based on the author's experience as a home inspector, and on the author's experience in helping to develop and maintain SOPs. Reasonable people may have different opinions.

Disclaimer of Liability

Every effort has been made to ensure the accuracy of the material contained in this book. ASHI, ASHI Education, Inc., the author, and the publisher assume no liability for any loss or damages caused by the use or interpretation of the material in this book. ASHI, ASHI Education, Inc., the author, and the publisher fully disclaim liability to all parties for all losses or damages including, but not limited to, incidental and consequential damages, and losses or damages caused by deficiencies, errors, or omissions contained in this book regardless of whether such deficiencies, errors, or omissions result from negligence, accident, or any other cause or theory.

ACKNOWLEDGEMENTS

My exposure to home inspection standards of practice began soon after I joined ASHI. John Cranor recruited me to be a member of the Standards Committee to which he had been appointed as the chair. Thank you John. After three years on the committee under John's leadership, Bill Richardson asked me to take over for John as committee chair. Thank you Bill. I remained as standards chair for six years until I was elected to the ASHI Board of Directors. Thank you all ASHI Presidents who reappointed me.

I am still involved with standards as the Board Liasion to the committee. Thank you Tom Lauhon, the current standards chair, and to all those who have served on this committee over the years. Your important work is seldom recognized or appreciated, but it is vital to what home inspectors do every day.

Thank you to Harry S. Rosenthal, Esq., ASHI's attorney, for his help in reviewing this book. I have become pretty good at channeling Harry over these many years, so for him, this disclaimer. Harry only reviewed this book for legal issues, not for technical issues.

Last, but by no means least, thank you Patricia, my wonderful wife. She edited this book and all of my other books and articles. She gives me good advice and she continues to be understanding about the signficant amount of time that I devote to ASHI.

Bruce A. Barker

Cary, North Carolina

July, 2018

CONTENTS

INTRODUCTION TO STANDARDS OF PRACTICE (SOP)

OBJECTIVES OF A STANDARD OF PRACTICE

This commentary section discusses why standards of practice (SOPs) exist. This discussion is not limited to the American Society of Home Inspectors (ASHI) Standard of Practice for Home Inspections (HISOP). It includes other home inspection SOPs, and SOPs for other services, such as predrywall inspections.

Going Beyond the Minimum

The first objective of a SOP is to establish minimum parameters for an inspection service. Minimum can have a negative connotation; however, it does not have a negative connotation when discussing a SOP. An inspection performed in compliance with a SOP provides the client with a full and complete service that satisfies reasonable expectations. Thus, it is not necessary to go beyond a SOP; however, many inspectors do so. An inspection that goes beyond a SOP can be a better inspection, **but this does not make a SOP-compliant inspection a deficient inspection**.

Inspectors often attempt to differentiate themselves in the marketplace by providing services and by using tools beyond those mandated in a SOP. These services and tools may be provided intentionally, and may be promoted through such means as advertising or verbal promises. These services and tools may be provided unintentionally by reporting about out-of-scope systems, components, or conditions that were observed while inspecting in-scope systems and components. Both attempts at differentiation are acceptable, if the inspector educates the client about realistic expectations.

Inspectors should be creative about providing services and using tools beyond those mandated in a SOP; however, inspectors should help the client have realistic expectations about these services and tools. Using tools such as infrared cameras and drones, and reporting about out-of-scope systems, components and conditions, could create unrealistic expectations.

Inspection agreements and reports should include a statement that using out-of-scope tools and reporting about out-of-scope systems, components, and conditions is done as a courtesy to the client, and does not alter or expand the inspection scope as stated in the SOP. Inspection agreements and reports should educate clients about the limitations of out-of-scope services and tools, such as infrared cameras, drones, and moisture meters. Inspectors should discuss out-of-scope services and tools with the client at the inspection.

Defining the Home Inspection Services

The second objective of a SOP is to define the agreement between the client and the home inspector about the services provided by the home inspector. A SOP defines the objective, scope, and limitations of the services. Without this agreement, misunderstandings about the services are likely. These misunderstandings may need to be resolved by a third party.

If a SOP does not exist for a specific service, the inspector should create an addendum to the inspection agreement that, at a minimum, educates the client about the limitations of the service. For example, limitations of using an infrared camera as part of an inspection include that weather conditions may limit the ability of the camera to detect defects, and that some defects do not produce enough of a temperature difference at the surface of an object to detect a defect below the surface. As is true for all inspection agreements, the inspector should seek legal advice about the addendum.

Protecting the Inspector

The third objective of a SOP is to help protect the inspector against unrealistic expectations. The home inspection profession is a risky profession. Even if the inspector performs a good inspection, there is a risk that the client may not be satisfied.

Guiding the Inspector

The fourth objective of a SOP is to provide general guidance to the inspector about:

(1) systems and components that the inspector is required to inspect,

(2) systems and components that the inspector is not required to inspect,

(3) conditions that may constitute a reportable defect or other reportable condition, and

(4) systems and components that the inspector should describe in the report.

General guidance is all that a properly trained inspector should need to conduct an inspection in compliance with a SOP. A SOP cannot provide specific guidance. Specific guidance would require compiling and updating information about hundreds of thousands of systems and components, and about their potential reportable defects. This is not practical, or necessary. Specific guidance is the domain of education for new inspectors, and continuing education for all inspectors.

HISOP ORGANIZATION AND CONTENTS

The HISOP is organized as follows. Sections contain major topics. For example, Section 2 describes the purpose and scope of a home inspection. Sections are indicated by whole numbers. Sections are divided into subsections. For example, Subsection 2.1 describes the purpose of the HISOP and the objective of an inspection. Subsections are indicated by a decimal number, such as 2.1. Subsections are divided into clauses. For example, Clause 2.2.A describes what the inspector should do during an inspection. Clauses are indicated by a decimal number, a capital letter, and sometimes another number, such as 2.2.B.1.

HISOP Section 1 is a brief history of ASHI and its purpose.

HISOP Section 2 (which may be the most important section in the HISOP) describes the purpose and scope of the HISOP, describes the objective of an inspection, presents the general requirements for inspection report contents, and identifies the conditions that constitute a reportable defect or condition. This section also commits inspectors to comply with the ASHI Code of Ethics for the Home Inspection Profession (COE). An inspection performed without compliance with the COE does not comply with the HISOP.

HISOP Sections 3 through 12 identify the systems and components that inspectors are required to inspect, and identifies specific systems and components that inspectors are not required to inspect. These sections also identify the systems and components that inspectors are required to describe in the inspection report.

General limitation clause 13.1.A is important when determining the systems, components, and conditions that inspectors must report using the HISOP. This clause states that a system, component, or condition is out-of-scope unless the system, component, or condition is specifically identified in Sections 2 through 12 as being in-scope. Thus, Sections 2 through 12 are not intended to, and do not, identify all out-of-scope systems, components and conditions.

HISOP Section 13 identifies the general limitations and exclusions that apply to an inspection performed using the HISOP. These limitations and exclusions apply to all HISOP sections.

HISOP Section 14 contains definitions of terms used in the HISOP. This is another very important section. It is more difficult to fully understand the HISOP without understanding these definitions. The definitions, when applied, are intended to clarify the meaning and intent of the defined terms. Terms printed in an italic font are defined terms in Section 14.

HISOP Section 1 - HISOP Introduction

INTRODUCTION

This section is a brief description of ASHI. It reads as follows.

The American Society of Home Inspectors®, Inc. (ASHI®) is a not-for-profit professional society established in 1976. Membership in ASHI is voluntary and its members are private home inspectors. ASHI's objectives include promotion of excellence within the profession and continual improvement of its members' inspection services to the public.

HISOP Section 2 - Inspection Purpose and Scope

INTRODUCTION

This section is probably the most important section in the American Society of Home Inspectors (ASHI) Standard of Practice for Home Inspections (HISOP). It states, among other things, the objective of an inspection, what inspectors must do during an inspection, conditions that inspectors must report, and contents that inspectors must include in the inspection report. All inspectors must have a thorough understanding of this section.

The scope of an inspection is defined as the systems, components, and actions that inspectors must include in the inspection. The HISOP identifies these systems, components, and actions by using the phrase: "The *inspector* shall." The HISOP identifies the systems, components, and actions that inspectors are not required to include in the inspection using the phrase: "The *inspector* is NOT required to." HISOP Section 13 contains a long list of general limitations and exclusions that inspectors are not required to include in the inspection.

Inspection scope is described in several sections of the HISOP. Systems, components, and actions that inspectors must include in the inspection are in-scope. Systems, components, and actions that inspectors are not required to include in the inspection are out-of-scope. Refer to specific HISOP sections, and to Section 13, when determining what is and is not in-scope.

2.1 – STANDARD OF PRACTICE VERSUS STANDARD OF CARE

This Subsection states that the HISOP establishes a minimum standard for performing an inspection. A minimum standard was previously discussed in the Objectives of a Standard of Practice section of this commentary. Inspectors are free to exceed the minimum standard (see also Clause 2.3.A), but the HISOP does not create an obligation to do so. In fact, Clause 13.1.A states that only actions, determinations, and recommendations specifically stated in the HISOP are required, everything else is out-of-scope.

While the HISOP does not obligate inspectors to exceed the minimum standard, a concept referred to as the standard of care may obligate inspectors to report about systems, components, or conditions, or to perform actions, that are not specifically required in the HISOP. The standard of care comes from negligence. It may be defined as what a reasonable inspector would do under similar conditions in the market where the inspector does business. For example, if most inspectors in a market use a combustible gas detector to inspect accessible gas fittings, then this procedure might be considered in-scope, even though the HISOP defines use of a combustible gas detector as a technically exhaustive (a defined term) procedure, and is therefore out-of-scope.

The standard of care can be common throughout the inspection profession, or it can be based on regional practices. Thus, what constitutes the standard of care can be difficult to determine. Inspector association local chapter meetings and local continuing education events are good places to learn about the local standard of care.

2.1 – INSPECTION OBJECTIVE

The objective of a home inspection is to provide clients with information about the condition of the **inspected** systems and components **at the time of the inspection**. This sentence contains two important scope limitations. First, only those systems and components that are actually inspected are in-scope. Systems and components that are not inspected, for whatever reason, are out-of-scope. Inspectors must notify the client in the report if in-scope systems and components were not inspected, and the reason why not (see also Clause 2.2.B.4). Second, inspectors can only report about conditions at the time of the inspection. Conditions change. Something that functioned properly during the inspection can cease functioning immediately after the inspection. This is an important concept that both the client and the inspector should understand.

2.2.A – INSPECTION REQUIREMENTS

This clause states the procedures an inspector must follow when conducting an inspection. Inspectors must inspect readily accessible, visually observable, installed systems and components listed in the HISOP.

Terms (such as *inspect* and *readily accessible*) that are italiczed in the HISOP are defined in Section 14. In order to understand inspection procedures, it is essential that inspectors understand the required procedures, the definitions of the italicized terms, and the requirement for probing.

It is important to remember that this document applies to the ASHI HISOP. Other SoPs, including SOPs in states that license home inspectors, may have different requirements.

Inspect and Home Inspection (defined terms)

An inspection consists of four tasks:

(1) a visual examination of in-scope systems and components identified in the HISOP that were present and readily accessible (a defined term) during the inspection,

(2) operation of systems and components specified in the HISOP using normal operating controls (a defined term),

(3) opening readily openable access panels (a defined term),

(4) production and delivery of a written inspection report that complies with the HISOP reporting requirements.

Readily Accessible (defined term)

A system or component is readily accessible when it is in plain sight and can be safely approached closely enough to allow visual observation. Inspectors are usually not required to, and in many cases should not, make an inaccessible system or component accessible.

A system or component is not readily accessible if it is concealed. Examples include underground systems, or components such as fuel tanks, water service pipes, building sewer pipes, electrical service laterals, and foundation walls and footings. Other examples include structural components such as walls, floors, and ceilings that are covered by finish materials (see also Clauses 13.1.B.2.a, 13.2.A.1, and 13.2.E.1).

The interiors of many systems and components are not readily accessible. Examples include the interiors of most chimneys, combustion vents, and plumbing pipes. The interior of wall cavities is not readily accessible.

A system or component is not readily accessible if in order to make the system or component accessible: (1) something, such as occupant belongings, must be moved, or (2) the system or component must be dismantled, (a defined term), or (3) property may be damaged, or (4) people's (including the inspector's) health or safety may be at risk (see also Clauses 3.2.C, 3.2.D, 13.2.D.1, 13.2.F.1, 13.2.F.3, and 13.2.F.4).

Many inspectors move some belongings and other material in order to gain access to a system or component, even though this is not required. Each inspector must determine what, if anything, to move based on conditions at the inspection, such as the type and quantity of belongings and materials to be moved, and the inspector's judgment about the risks involved (see also Clause 13.2.F.3). An inspector may reach a different conclusion about whether or not to move similar belongings and materials during the same inspection, and may reach a different conclusion at different inspections. An inspector should attempt to have a consistent policy about moving belongings and materials; however, absolute consistency is not required. Each situation is different.

An inspector should exercise care when moving things. The drug store trinket becomes a priceless heirloom when broken.

It is prudent to take a picture to document situations in which a system or component is not readily accessible. The picture should be retained with the inspection records in case questions arise about why a system or component was not inspected.

Readily Openable Access Panel (defined term)

Some in-scope components are located behind covers or panels that inspectors must open or remove to gain access. Inspectors often encounter two types of covers or panels. One type is found on manufactured appliances, such as furnaces. The other type provides access to crawlspaces and attics. Inspectors should consider three factors to determine whether inspectors are required to open or remove a cover or a panel.

The first factor is whether the cover or the panel is intended for the homeowner to open or to remove for inspection, operation, or maintenance. Manufacturer's instructions usually define whether there are homeowner-serviceable parts inside of manufactured appliances. Crawlspace and attic access openings are presumed to be intended for homeowner use.

Since it is not practical, or required, to read manufacturer's instructions (see also Clause 13.2.A.8), common sense dictates an inspector's decision whether to open or to remove a cover or a panel. For example, if the cover or panel grants access to a gas valve, pilot light, or a filter that the homeowner must operate to use the appliance, inspectors usually should open or remove the cover or the panel. Low and medium efficiency gas furnaces and gas fireplaces are common examples of this situation. Examples of covers or panels that inspectors are usually not required to open or to remove include those on air conditioning condensers and evaporator coils, heat pump air handlers, and ground-mounted and roof-mounted heating and air conditioning package systems.

The second factor is cover or panel accessibility. The cover or panel must be readily accessible. It must be within normal reach. Some define normal reach as requiring a short ladder or stool. Others define normal reach as not requiring a ladder or stool. Normal reach is an example of something that might be defined by local inspection practices.

The third factor is whether the cover or panel is sealed. Whether a cover or panel is sealed depends on the method of sealing. If the sealing method would be damaged by opening or removing the cover or panel, then the panel is sealed, and inspectors are not required to open or remove it. Examples sealing methods that would be damaged include: tape, mastic, coatings such as paint, caulk, and drywall joint compound. Fasteners such as screws do not necessarily make a panel sealed. A common inspection practice is that inspectors will remove up to four fasteners, if the panel is intended to provide access to the homeowner or occupant .

Inspectors are not required to open or remove a panel if the inspector believes that doing so may cause damage, or if the inspector believes that it may not be possible to place the panel back into its original position (see also Clause 13.2.F.1). Inspectors should report if a readily openable access panel was not opened or removed, report the reason why, and recommend further evaluation of the appliance (see also Clause 2.2.B.4).

Conditions are different at each inspection. An inspector may reach a different conclusion about whether or not to open an access cover or panel at each inspection, depending on conditions at the inspection. An inspector should attempt to have a reasonable and consistent policy about opening access covers and panels.

Dismantle (defined term)

Inspectors are not required to, and in most cases should not, dismantle any system or component, unless the homeowner would do so during normal maintenance of the system or component. To dismantle a system or component, in the context of an inspection, is to disassemble a system or a component into constituent parts, or to remove a component to gain access to another component. Dismantling usually requires the use of tools. Examples of dismantling that inspectors are not required to perform include: removing furnace parts to gain access to the heat exchanger, removing sections of a vent connector to gain access to the vent connector or vent interior, and removing fasteners that hold an appliance in place to gain access to concealed parts of the appliance.

Removing the dead front cover from an electrical panel enclosure is an example of dismantling. This procedure, however, is specifically required by Clause 7.1.A.5, and inspectors are, therefore, required to remove the dead front cover from an electrical panel enclosure (see also Clause 13.2.F.4).

Dismantling and opening or removing readily openable access panels are different procedures. Inspectors should not apply the criteria for one procedure to the other procedure.

Installed (defined term)

A system or component is considered installed if it is attached to the property in a way that removal requires tools. Systems and components that are often sold with the house are frequently considered installed. For example, a microwave oven mounted over a range is considered installed. That same microwave oven located on a countertop would not be considered installed, but it would be considered installed if it were attached in an alcove below the countertop.

Normal Operating Controls (defined term)

Normal operating controls are intended to be operated by the homeowner during normal daily activities. These controls include thermostats, light switches, and faucets. They do not include plumbing valves such as stop valves and shutoff valves, and electrical devices such as equipment disconnect switches and circuit breakers (see also Clauses 13.2.C.3 and 13.2.C.4).

Probing (Clause 13.2.F.7)

Probing is a procedure wherein a sharp object, such as an awl or an ice pick, is inserted into wood to determine if the wood is damaged or deteriorated. Probing is required only when the inspector has visible evidence that deterioration may exist, and when probing will not damage the surface being probed. Damage is likely to occur when probing a finished surface.

Probing is often not required, and is often not necessary. If visible deterioration exists, probing may help determine the extent of the deterioration, but determining this is out-of-scope. Probing is not required if visible deterioration does not exist, and the inspector has no reason believe that deterioration exists.

System or Component Location

Whether a system or component is in-scope depends on where the system or component is located (see also Clauses 13.1.B.2.a, 13.1.C, 13.2.A.1, and 13.2.E.5). Systems and components are usually:

1. in-scope if they are located inside, or if they are attached to the building being inspected.

2. out-of-scope if they are located inside, or if they are attached to buildings that are not being inspected.

3. out-of-scope if they outside, and are not attached to the building being inspected.

4. out-of-scope if they serve out-of-scope systems, such as swimming pools and spas.

Some inspectors inspect out-of-scope systems or components, but this is not required.

2.2.B – REPORTING REQUIREMENTS

This clause tells inspectors about four of the five required elements that must be a part of every inspection report. The fifth required element is to describe systems and components identified in Sections 3 through 12. An inspection is not complete, and does not comply with the HISOP, unless all of the required elements are present in the report.

Report Objective

While not specifically stated, the report should present information in a manner that a client can understand and use. The requirement to present information in an understandable manner is implied Subsection 2.1 and in clauses such as 2.2.B.2 and 2.2.B.3. Reports that a client can understand and use avoid jargon, technical terms, uncommon words, contradictory statements, improper spelling and grammar, and improper or incorrect descriptions and terms.

Report Format

An inspection report must be written. Pictures, illustrations, video, and audio may be included, and may supplement a written report; but these may not replace a written report.

Inspectors may select the report medium. Handwritten reports are allowed, but few inspectors use them because in many markets clients and real estate agents will not accept them. Computer-generated reports are the modern medium.

Inspectors may select the report format. Checklist-style reports are allowed if they comply with all reporting requirements, which they sometimes do not. A checklist-style report is a report that contains a list of common defects and descriptions. The inspector places a mark on the report to indicate that a defect is present, and places a mark on the report to describe systems and components. The inspector usually adds notes to report defects, and to describe systems and components that are not included on the checklist.

Narrative-style reports are allowed. A narrative-style report describes systems and components and reports defects using words. The inspector usually draws narrative-style report content from a library of pre-written narratives.

Hybrid reports are allowed. Hybrid reports combine elements of checklist and narrative reports.

Some inspectors include a summary report in addition to the full report. A summary report typically lists only the reported defects. Some report software also generates a repair addendum that the client's agent may use. These summary reports and repair addenda should contain a statement warning the client that these partial reports do not contain all of the important information that the client needs to make decisions about the properly, and should advise the client to read the full inspection report before making decisions.

Deficiency Statements

A building without at least one deficiency probably does not exist. Most inspection reports will, therefore, have at least one deficiency statement. Every deficiency statement must have the elements specified in Clauses 2.2.B.1 and 2.2.B.2, and most deficiency statements must have the element specified in Clause 2.2.B.3. These elements are discussed in detail in the following pages. A method for remembering these elements is that a deficiency statement should:

(1) describe the deficiency observed, and

(2) explain why the client should care about the deficiency, and

(3) recommend the action that the client should take to address the deficiency.

Affirmative Reporting

Affirmative reporting occurs when inspectors report the condition of all inspected systems and components. Inspectors should report the condition of systems and components that do not have reportable deficiencies using terms such as acceptable and serviceable. It does not matter what term inspectors use when reporting the condition of systems and components without reportable defects, if the inspector adequately defines the term in the report.

Inspectors should report the condition of systems and components with reportable deficiencies using a deficiency statement. Inspectors are required to use the deficiency statement elements contained in Clauses 2.2.B.1, 2.2.B.2, and 2.2.B.3.

Affirmative reporting is not required by the HISOP. Inspectors are only required to report about systems and components with reportable deficiencies. Affirmative reporting is required by some other standards of practice.

2.2.B.1 – CONDITIONS THAT MUST BE REPORTED

This clause describes conditions that inspectors must identify during the inspection and include in the report. While conducting the inspection, inspectors must make a professional judgment about whether a condition should be reported, and if so, inspectors must make an additional professional judgment about how to provide the client with information about the condition. Inspectors must make these professional judgments many times during an inspection. These professional judgments are based in part on training and education, and in part on experience. Inspectors should actively participate in as much good quality training and education as possible in order to make good professional judgments about identifying and reporting these required conditions.

It is important to distinguish between functional deficiencies and cosmetic deficiencies. Cosmetic deficiencies are those that do not significantly affect a component's performance of its intended function. Inspectors are not required to inspect for or to report cosmetic deficiencies (see also clause 13.1.B.2.b).

Not Functioning Properly

Inspectors must identify and report in-scope systems and components that are not functioning properly. Other standards of practice use similar expressions of this requirement such as not functioning as intended, and not functioning as designed. The objective is the same regardless of the term used.

Not functioning properly is a condition that is present and observable during the inspection. This condition often applies to heating/air conditioning, plumbing, and electrical systems and components, but it can apply to other systems and components. This condition can present as an obvious failure to function at all, or as a more subtle malfunction such as a yellow gas flame, or a red oil flame. Significant visible deformation of structural components could be described as not functioning properly because they are not supporting the intended load. Low water flow from a plumbing fixture often means that the fixture or pipe serving the fixture is not functioning properly.

Significantly Deficient

Significantly deficient is also a condition that is present and observable during the inspection. It is possible, however, that a system or component is functioning properly during the inspection, and that it is also significantly deficient (examples follow). The difference between not functioning properly and significantly deficient is that a significantly deficient system or component lacks an essential quality or element that could prevent it from functioning properly under certain conditions, or at some future time. A significantly deficient system or component may function properly for many years, even decades. It may never malfunction.

There are many examples of significantly deficient conditions that present no evidence of a malfunction during the inspection. A joist or beam that has an excessively large notch or hole may be supporting the load imposed when inspected, but it may deform or fail if a larger load is applied. A wire that is a smaller gauge than is allowed for the circuit breaker to which the wire is connected may present no visible evidence of a malfunction when inspected. Concealed damage could exist elsewhere in the circuit, or the cumulative effects of excess heat in the wire may not have occurred, yet. Wood that is too close to a fireplace or chimney may present no visible indication of a defect for decades; however, pyrolysis may be slowly changing the molecular structure of the wood so that it may ignite during the next fire in the fireplace.

Identifying a significantly deficient condition can be difficult. Damage, deterioration, and deformation can be obvious visible clues that a significantly deficient condition exists. Often, however, inspectors must draw upon education and experience to identify more subtle clues that significantly deficient conditions exist. Examples of possible significantly deficient conditions include: failure to comply with building codes, failure to comply with manufacturer's instructions, or failure to comply with industry standards; however, Clause 13.2.A.8

specifically excludes these from inspection scope. Inspectors may use building codes, manufacturer's instructions, and industry standards to identify significantly deficient conditions, but this is not required.

Significantly deficient presents a problem for inspectors. The HISOP requires identification and reporting of this condition; however, many of the means of identifying this condition are out-of-scope. The best resolution of this problem is good quality training and education.

Unsafe

Unsafe (a defined term) is a condition that is present and observable during the inspection. A system or component may be considered unsafe by inspectors if both of the following risks are present: (1) a significant risk of serious bodily injury exists, and (2) the risk exists during normal day-to-day use.

Unsafe only applies to risks that are reasonably likely to occur. It is not possible to assign a specific probability or time period to this risk. Unsafe may be defined by a time period measured in years between injuries, but probably not in a time period measured in decades.

Unsafe only applies to bodily injury that may be sustained by people. Unsafe does not apply to property damage, or deterioration of systems and components. The bodily injury must be significant. An injury requiring immediate medical treatment is one definition of a significant injury.

Unsafe applies when the system or component is used as it is designed, and when it is used for its intended purpose. Unsafe does not apply to misuse and abuse, but it may apply to accidents, and to reasonably foreseeable unintended uses.

The definition of what constitutes an unsafe condition changes over time. A condition that was considered safe and acceptable in the past may not be considered safe in the present. An inspector may consider a condition to be unsafe regardless of whether the condition complied with building codes or generally accepted practices when the system or component was installed. There is no grandfathering of unsafe conditions. An inspector may consider a condition to be unsafe even if it complies with the current building codes and standards.

What constitutes an unsafe condition is solely at the discretion of the inspector based on the definition of unsafe, and on conditions at the property being inspected. Reasonable inspectors can reach different conclusions, and the same inspector may reach a different conclusion based on different property conditions.

Unsafe conditions may also be considered significant deficiencies in some cases. For example, a deck ledger that is attached to the building using only nails is both significantly deficient and unsafe. Inspectors should report both the structural implication (the deck may pull away from the house), and the safety implication (people could be seriously injured). Reporting both implications helps the client understand the importance of correcting such deficiencies. Explaining deficiency implications is discussed further in Clause 2.2.B.3.

Absence of safety glazing is an example of a situation that can result in different conclusions about whether, and how, to report this as an unsafe condition. Safety glazing requirements change with almost every code revision cycle. For example, requirements for safety glazing in and around doors and near water have remained relatively consistent, but have changed. Thus, inspectors may be more likely to report lack of safety glazing in and around doors and near water. Inspectors may, however, consider the degree of risk, especially in older houses. The risk posed by lack of safety glazing in a ten-inch wide sidelight next to a door is much less than lack of safety glazing in a door made mostly of glass.

<u>Near End of Service Life</u>

The requirement to report systems and components that are near the end of their service lives presents another problem for inspectors because Clause 13.2.A.6 excludes determining future conditions, including the failure of systems and components. The problem is resolved by realizing that the requirement is not to determine, or to speculate about, when a system or component might fail. The requirement is to use available information to estimate the expected service life of a system or a component, and to evaluate the condition of the system or component with respect to the expected service life.

Note that the age of the system or component was not mentioned in the previous paragraph. Inspectors are not required to report the age of systems and components. While age is often a factor when determining whether a system or component may be near the end of its service life, it is not the only factor, and it is not always the most important factor. High quality systems and components that are well-maintained can last many years beyond the expected service life of similar systems and components. Conversely, lower quality systems and components, those that are poorly maintained, and those that are subjected to harsh conditions, can fail much sooner than similar systems and components.

Identifying systems and components that may be near the end of their service lives is subjective, since it is not based exclusively on the estimated age of the system or component. This identification is more of an art than a science, and involves experience as much or more than training. This is especially true for components such as roof coverings, for which age cannot be objectively determined, and for which condition is more important than age.

Inspectors should exercise care when reporting about systems and components that may be near the end of their service lives. Reports should avoid wording that states or implies a remaining service life. An end of service life statement may use wording such as: "The water heater appears near the end of its service life. These water heaters can fail at any time. We recommend monitoring the water heater for indications of leaking, and other indications of failure, and we recommend budgeting for replacement in the near future."

Descriptions – Systems and Components

Several HISOP sections require that inspectors describe certain systems and components. Describe is a defined term. The requirement is to describe systems and components so that the reader can distinguish the system and component from similar systems and components. Inspectors are not required to describe systems and components in detail. Accuracy is required, and inaccurate descriptions can result in dissatisfied clients.

There are many examples where a couple of missing words, or an inaccurate use of a term, can make a significant difference. Natural stone and artificial stone wall coverings are good examples of where precise wording is required. Both wall coverings appear similar, but natural stone is more like brick veneer, and artificial stone is more like stucco. Describing either as stone is incorrect, and could lead to a significant problem for inspectors. The better descriptions are natural stone veneer, and adhered masonry veneer, respectively.

Plywood panel wall covering is a good example of where inaccurate use of a term can make a significant difference. This wall covering is often described as T1-11. There are actually many profiles of plywood panel wall covering. T1-11 describes only one profile in which the raised area is eleven-inches wide and the recessed area is one-inch wide. To avoid confusion when replacing deteriorated siding, the more accurate description in this case is simply plywood panel wall covering; let the contractor determine the exact profile, if necessary.

Descriptions – Attic and Crawlspace Inspection Methods

Clause 3.1.B.1 requires inspectors to describe the methods used to inspect crawlspaces and attics. Most attics, and many crawlspaces, have areas that are not readily accessible, or that the inspector considers too dangerous to enter (see also Clauses 13.2.A.1, 13.2.D.1, 13.2.D.2, and 13.2.F.1). It is important, therefore, to describe the crawlspace and attic inspection methods, to describe the areas that were not inspected, and to explain the reason why they were not inspected (see also Clause 2.2.B.4).

Four situations exist when describing attic and crawlspace inspection methods: (1) not present, (2) not accessible, (3) partially accessible, and (4) fully accessible. Each situation requires a different description in the report. Often, multiple inspection methods are used. Inspectors should describe the method used to inspect each attic and crawlspace area.

It is not necessary to describe crawlspace inspection methods if a crawlspace is not present, such as houses with basement and slab foundations. Lack of a crawlspace in these houses should be self-evident to the average client.

Most houses have an attic, so the fact that some houses do not have an accessible attic may not be self-evident to the average client. Houses with low slope roofs, and houses with vaulted ceilings, do not have an accessible attic, or have a combination of accessible and inaccessible attics.

Inspectors should report when part or all of an attic is not accessible. If none of an attic is accessible, inspectors should report this fact. If some parts of an attic are accessible and some parts are not accessible, inspectors should describe which parts are accessible and which parts are not accessible. Describing the accessible and inaccessible parts by their location above identifiable rooms is a good method that should be clear to the average client. An inaccessible attic area above a family room with a vaulted ceiling may be described use wording such as: "There is no accessible attic above the family room. This area was not inspected, and defects in this area, if any, were not identified and reported."

Many accessible attics and crawlspaces have areas that are not accessible, or that may be viewed only from a distance. In these situations, inspectors should be specific about which areas were not accessible, and about which areas were viewed from a distance. Describing these areas by their location, and by their approximate percentage of the attic or crawlspace area, is a good method that should be clear to the average client. For example, most accessible attics have low clearances near the eaves that make these parts of the attic visible only from a distance. It is prudent to report that defects in these areas may not be visible. An inaccessible crawlspace area may be described use wording such as: "Approximately twenty-five percent of the right rear crawlspace area was not accessible because of low clearance, and was viewed from a distance. Defects in this area, if any, were not identified and reported." Inspectors may wish to provide a couple of examples of defects that may not be identified and reported; however, this is not required. Inspectors may use similar wording to describe an inaccessible attic area.

Descriptions – Roof Inspection Methods

Clause 5.1.B.2 requires inspectors to describe the methods used to inspect the roof covering and related components such as gutters, flashing, and roof penetrations. Many roofs have areas that are not readily accessible, or that the inspector considers too dangerous to access (see also Clauses 13.2.A.1, 13.2.D.1, 13.2.D.2, and 13.2.F.1). It is important, therefore, to describe the roof inspection methods, to describe the roof areas that were not inspected, and to explain why they were not inspected (see also Clause 2.2.B.4).

Four situations exist when describing roof inspection methods: (1) roof walked upon, (2) roof viewed from ladder, (3) roof viewed from ground, and (4) roof not visible. Each situation requires a different entry in the report. Sometimes multiple inspection methods are used. Inspectors should describe the inspection method used to inspect each roof area.

Inspectors are not required to walk on the roof per Clause 13.2.F.1, even if the roof is readily accessible. Many inspectors walk on readily accessible roofs because this is the best way to inspect a roof; however, each inspector must determine whether or not the inspector considers the roof safe to walk on. The inspector should report if he/she was able to walk on the entire roof.

The next best roof inspection method is to view the roof from the edge while remaining on the ladder. Inspectors should use this method to view the roof from as many vantage points as safely possible, if the inspector elects not to walk on the roof.

The HISOP does not prescribe a minimum or maximum ladder height. Ladder height may be influenced by local practices, and by competitive considerations. Inspectors may wish to consider using a ladder similar to that used by other inspectors in the market; however, each inspector should determine the type and height of ladder that he/she believes is safe for him/her to use regardless of other considerations.

Roofs that are too high, too steep, too wet, covered by snow, or that are otherwise not safely accessible from a ladder may be inspected from the ground. Many inspectors use binoculars when inspecting the roof from the ground; however, this is not required by the HISOP. A ground-level roof inspection is a limited inspection that may not allow identification of defects, and inspectors should report this limitation to the client.

Parts of the roof, or the entire roof, may not be visible by any means. Sometimes the roof is covered by snow. Sometimes it is not possible to see parts of the roof from the ground, or from any vantage point. Inspectors should report this limitation to the client using wording such as: "The left side of the front gable roof was not inspected because it was not possible to view this area from any vantage point. Defects in this area, if any, were not identified and reported."

2.2.B.2 – RECOMMENDATION TO CLIENT

This clause requires inspectors to recommend an action that the client should take to address a deficiency reported per Clause 2.2.B.1. Every deficiency statement must contain a recommendation to address a reported deficiency, including systems and components that are near the end of their service lives.

This clause provides three options that inspectors may recommend to clients: correct, monitor, or further evaluation (a defined term). Inspectors should recommend one of these options in every deficiency statement.

Inspectors are not required to, and in most cases should not, specify or design a specific action to correct the deficiency (see also Clauses 13.2.A.5, 13.2.B.2, and 13.2.B.3). Inspectors may specify or design a specific action (see also Clause 2.3.B), however, inspectors assume additional risk by doing so. Specifying or designing specific actions often requires additional information that is not available to inspectors, and sometimes the inspector does not possess the knowledge or training to specify or design specific actions.

An example of a specific action is recommending that a deficient system or component be repaired, or that it be replaced. Inspectors should avoid recommending only one of these options. Inspectors may recommend that a specialist consider both options, and recommend the appropriate option to the client. Repair usually costs less than replacement; therefore, recommending repair may cause the client to incorrectly interpret the deficiency as being minor when in fact a costly replacement may be the appropriate method of correcting the deficiency.

Correct the Deficiency

Correction is the preferred recommendation for conditions reported as not functioning properly and reported as significantly deficient. Inspectors should have a valid reason for using other client action recommendations when reporting these deficiencies. A correction recommendation may use wording such as: "We recommend repair or replacement as recommended by a qualified contractor." Inspectors may recommend a specific professional, such as an engineer, depending on the situation. This topic is discussed in more detail later in this chapter.

A spliced electrical wire that is not in a covered box is a good example of a situation where correction is the appropriate recommendation. The condition is clearly significantly deficient, the correction is usually simple, and replacement is usually not required. In this case, recommending repair by a qualified electrician is appropriate.

Lack of functional drainage in a plumbing fixture is another a good example of a situation where correction is the appropriate recommendation. The system is clearly not functioning properly, and the correction is usually, but not always, simple. In this case, recommending repair may not be appropriate. Recommending action by a qualified plumber is appropriate. The plumber decides what action, such as repair or replacement, is appropriate.

Further Evaluation

Further evaluation is a defined term. The definition is: "examination and analysis by a qualified professional, tradesman, or service technician beyond that provided by a home inspector." Further evaluation is most appropriate when the inspector observes evidence of a deficiency, but confirming the existence, location, or nature of the deficiency is either not possible, or is out-of-scope. A further evaluation recommendation may use wording such as: "We recommend further evaluation by a qualified contractor, and action, if necessary, as recommended by the contractor."

A water stain is a good example of a situation where further evaluation is the appropriate recommendation. The stain itself is not the deficiency, it is only evidence that a deficiency may exist. The actual deficiency is the source of the moisture causing the stain. Even if the inspector can determine that the deficiency is active (the stain is wet), or even if the inspector can determine the existence, location, or nature of the moisture source, further evaluation is often necessary to confirm the inspector's findings, and to determine the actions required to correct the deficiency.

A low temperature difference in an air conditioning system is a another good example of a situation where further evaluation is the appropriate recommendation. A low temperature difference may or may not be evidence that the system is not functioning properly. Further evaluation by a qualified HVAC technician is necessary to determine if there is a deficiency, and if so, to determine the actions required to correct the deficiency.

Recommending further evaluation is appropriate when the inspector is confronted with a system or component with which the inspector is unfamiliar, or with which the inspector has limited knowledge or experience. This situation happens even to the most experienced inspectors. In this situation, it is not only appropriate, but it is also recommended, that the inspector report his/her inability to properly inspect the system or component, and to recommend evaluation by someone who has the skills to perform a proper inspection. It is better to admit this inability than to perform an incompetent inspection, and miss a major deficiency.

Further evaluation is sometimes improperly used by inspectors in an attempt to reduce risk to themselves, or to transfer risk to the client. Some inspectors recommend further evaluation of almost everything, prompting the reasonable question: "Why did I hire you to tell me to have all these other people evaluate everything?" If the existence of a deficiency is clear, then correction is the appropriate recommendation. Evaluation will often be required to determine the nature of the correction, but correction is still the appropriate recommendation.

Monitor

Monitor is a recommendation that inspectors should use in limited situations. Monitor may be appropriate when reporting systems and components that are near the end of their service lives. Monitor may also be appropriate when the inspector is uncertain whether a deficiency exists, and if one exists, the inspector believes that the deficiency has not risen to the level that requires immediate correction or further evaluation.

A hairline crack of uniform width in the mortar of a concrete block foundation wall or a brick veneer wall covering is a good example of when monitoring may be the appropriate recommendation. These cracks have multiple possible causes, many of which are benign and do not require immediate correction or further evaluation.

A monitor recommendation may use wording such as: "We recommend that you monitor the crack. If the crack increases in width or length, or if there is evidence of water infiltration through the crack, we recommend evaluation by a qualified contractor." Note that this recommendation provides the client with parameters to look for when monitoring the crack. A monitor recommendation may not provide the client with useful information without these parameters. Providing these parameters is not required, but it is recommended when appropriate.

Monitor is sometimes improperly used by inspectors to avoid being labeled as an alarmist. Monitor should not be used when a deficiency exists that should be corrected or evaluated. If uncertain, inspectors should recommend correction or evaluation.

Caution

Caution is not one of the recommendations cited in the HISOP. It can, however, be an appropriate recommendation when addressing situations such as those involving marginally unsafe conditions, especially when the cost to correct the condition is high relative to the risk. Minor trip hazards in walkways and driveways are a good example of where caution may be an appropriate recommendation. A caution recommendation for a minor walkway trip hazard may use wording such as: "We recommend exercising caution when using the walkway, and we recommend that you consider installing supplemental lighting to increase safety at night."

Type of Specialist

Recommending the type of specialist that the client should consult is not required by the HISOP; however, many inspectors do so. Doing so is prudent, and is consistent with the objective of providing the client with useful information. In many cases, the type of specialist is obvious. Recommend a plumber for plumbing and fuel distribution deficiencies, an electrician for electrical deficiencies, and an HVAC contractor for HVAC deficiencies.

Structural deficiencies can present a dilemma about whether to recommend a contractor or an engineer. Deficiencies involving trusses and engineered wood products such as I-joists should be referred to an engineer in almost all cases. Deficiencies for which there are commonly accepted repairs, manufacturer's instructions, or prescriptive methods in building codes may be referred to a contractor. Other deficiencies may be referred to a contractor based on the theory that a qualified contractor will consult an engineer if the contractor believes the repair is outside the contractor's allowed scope of work.

Recommending an engineer, or any other specialist, often means an additional expense for the client, and as such inspectors are sometimes reluctant to recommend an engineer. Inspectors should not consider cost when deciding to recommend an engineer or other specialist. It is usually better to spend a little to get an informed second opinion, than to risk not addressing a major defect.

Inspectors are often asked to recommend a specific specialist. Inspectors may do this, but should be very cautious. Inspectors should not accept any type of compensation for recommending a specific specialist, whether the compensation is direct, such as money, or indirect, such as gifts or discounted work (see also Code of Ethics Clause 1.F). Inspectors should remember that their reputations are vulnerable when recommending specific specialists, so screening recommended specialists is very important.

2.2.B.3 – EXPLAINING THE DEFICIENCY IMPLICATION

This clause requires inspectors to explain in the report the implication of deficiencies reported per Clause 2.2.B.1, including deficiencies reported as unsafe or near the end of service life. The exception involves deficiencies that are "self-evident."

These explanations are an important part of providing information that clients can use, and are an important part of reducing inspector risk. These explanations are, unfortunately, often absent in inspection reports. It is recommended that inspectors provide an explanation of the implication for almost all deficiencies. Failure to provide an explanation where one is required means that the inspector is not complying with the HISOP. The inspector could be found at fault in this situation, even if the inspector identified and otherwise properly reported a deficiency.

These explanations need not be long and complicated. In fact, short and simple is usually better for client understanding. One simple sentence will suffice for many deficiency explanations. Two sentences may be required for a few complicated deficiencies, such as those involving structural deficiencies.

One-sentence explanations should be the norm for most simple deficiencies. The explanation of many electrical deficiencies can use wording such as: "This is a fire hazard and a shock hazard." The explanation of many fuel distribution system deficiencies can use wording such as: "This is a fire hazard and an explosion hazard." The explanation of most water-related deficiencies can use wording such as: "Water can damage the house and provide moisture for fungal growth." Most clients should understand these explanations.

The explanation of structural deficiencies often depends on the type of deficiency, such as: not functioning properly, or significantly deficient. Structural components that are not functioning properly will often present visible deformation, deterioration, or significant cracking. The explanation of these deficiencies can use wording such as: "Further deformation or damage may occur, resulting in costly repairs." The explanation of deficiencies involving structural components that are significantly deficient, but functioning properly during the inspection, can use wording such as: "No evidence of deformation was observed; however, deformation or failure could occur if conditions change."

2.2.B.4 – REPORTING SYSTEMS AND COMPONENTS NOT INSPECTED

This clause requires inspectors to report systems and components identified in HISOP Sections 3 – 12 that were present during the inspection and not inspected, and to report the reason why they were not inspected. Inspectors should report these limitations regardless of whether it is required by this clause. Doing so helps avoid misunderstandings about the limitations of the inspection, and provides clients with information about additional inspections that may be required.

An inspector's ability to visually inspect systems and components is usually limited in completed structures. Wall structural components are usually concealed by finish materials such as drywall. Ceiling structural components are usually concealed by finish materials below, and by insulation above. Floor structural components are often at least partially concealed by insulation, and are usually fully concealed by finish materials on floors and on ceilings. Concealed systems and components are out-of-scope (see also Clauses 13.1.B.2, 13.2.A.1, and see also the definition of readily accessible).

Inspectors' ability to visually inspect systems and components is often limited in occupied structures because of occupant belongings, and sometimes because of locked doors. Inspectors should consider such systems and components concealed or not readily accessible, and therefore out-of-scope, because attempting to look around or behind belongings may not allow a complete inspection, and may produce an incorrect conclusion about the condition of the system or component. The same is true of systems and components behind locked doors.

Situations exist in which an in-scope system is inspected, but important components of the system are not inspected because they are concealed or are not readily accessible. Inspectors should consider reporting that these concealed or not readily accessible components were not inspected. Reporting this situation is not required because concealed and not readily accessible components are out-of-scope per clauses such as 13.1.B.2, 13.2.A.1, and 13.2.E.1. Reporting is prudent, however, because it avoids misunderstandings. Examples of such components that inspectors may wish to report may include: deck posts and footings located below grade, components concealed by suspended ceilings, and underground fuel storage tanks and piping.

Inspectors should also consider reporting the existence of out-of-scope systems and components that were present and that were not inspected. This is not required, but it helps avoid misunderstandings, and provides clients with useful information. It also helps reduce inspector risk. Examples of such systems and components include: fences, detached structures such as storage sheds, water features, fire pits and exterior fireplaces, play structures, sports structures, swimming pools and spas, grills and other exterior cooking appliances, landscape irrigation systems, radon mitigation systems, and common systems and components of multi-family housing buildings.

Most inspections occur in completed and occupied structures, and most of these structures present systems and components that are concealed, or that are not readily accessible. This presents two limitation reporting situations. One situation involves common conditions, such as interior and exterior wall coverings that conceal wall structural components, and the usual quantity of occupant belongings. The other situation involves uncommon or excessive conditions, such as occupant belongings or plants that conceal large areas.

General limitation statements should suffice for common conditions. For example, inspectors might place a statement near the wall structural component description that the wall structural components were concealed by finish materials. For example, inspectors might place a statement near the beginning of the report that occupant belongs concealed parts of the walls and floors.

Inspectors should use specific limitation statements when uncommon conditions exist, or when the inspector believes that it is prudent to alert the client that a limitation presents a significant and a likely risk that a defect may not be detected because of the limitation. For example, inspectors may report when boxes, other belongings, or plants completely obscure large areas of walls or floors. Specific limitations are also discussed in Subsection 2.2.B.1.

A limitation statement should contain four elements. The first two elements are required. The second two elements are not required, but are strongly recommended. These elements are: (1) identify the system or component not inspected in a manner that helps the client understand what was not inspected, (2) state the reason why the system or component was not inspected, (3) inform the client that the system or components could have deficiencies, and (4) recommend that the system or component be inspected, even if there is no practical way to do so.

2.2.C – ADHERENCE TO ASHI CODE OF ETHICS

This clause requires that inspectors who claim to comply with the HISOP also comply with the ASHI Code of Ethics for the Home Inspection Profession (COE). The COE is printed on the back of the pamphlet containing the HISOP. Compliance with the COE is an important part of complying with the HISOP. Commentary about the COE is beyond the scope of this document. ASHI members who wish an interpretation of the COE may submit a Request for Interpretation to ASHI Headquarters.

2.3.A – ADDITIONAL INSPECTION SERVICES

This clause allows inspectors to include whatever additional services that inspectors wish to offer in addition to the HISOP requirements. These services may include inspecting out-of-scope systems and components and providing services such as radon testing, water quality testing, lead paint sampling, air quality and mold testing, and inspections using drones and infrared cameras.

As previously discussed in the section about standard of practice objectives, inspectors should help the client have realistic expectations about such services. Inspectors should use appropriate limitation statements in the inspection agreement and in the inspection report when providing additional services without charge as part of the inspection. Inspectors should use inspection agreement addendums when providing additional services for additional fees.

Inspectors should be aware that some additional services require additional training, and some require licensing. Inspectors should not provide additional services without the necessary training, experience, and license.

2.3.B – DESIGNING OR SPECIFYING REPAIRS

This clause allows inspectors to recommend, specify, or design repairs if the inspector wishes to do so, and is qualified and licensed to do so. As previously discussed in the recommendation to clients section, specifying and designing repairs is not recommended because it increases inspector risk. This risk increase is due, in part, to the fact that the information necessary to make such recommendations is often not available to inspectors, at least without additional out-of-scope work. This risk increase is also due, in part, to the fact that many inspectors are not qualified and/or licensed to make such recommendations.

Inspectors should be aware that recommending, specifying, or designing repairs of deficiencies found during an inspection could violate the Code of Ethics, if these are done for additional compensation.

2.3.C – EXCLUDING REQUIRED INSPECTION SERVICES

This clause allows inspectors to exclude systems and components required by the HISOP, but only if the client requests exclusion, or if the client agrees to the exclusion. It is prudent to obtain this agreement in writing. Inspectors should report that such systems and components were excluded, in the same manner as with any other systems and components which were not inspected per Clause 2.2.B.4. The reason for not inspecting is client request.

HISOP Section 3 - Structural

INTRODUCTION

This section contains the requirements and specific limitations for inspecting the building structural components. The structural components that must be inspected are listed in Clause 3.1.B. These components are the foundation, floor, wall, ceiling, and roof.

3.1.A – STRUCTURAL INSPECTION REQUIREMENTS

The objective of the structural component inspection is to identify and to report visual evidence that a structural component may not be functioning properly, or that it may be significantly deficient. These components rarely create an unsafe condition, but if one exists, the inspector should report the unsafe condition, usually as an implication of the improper functioning or significant deficiency that is causing the unsafe condition (see also Clause 2.2.1.B.1, unsafe). The service life of these components, if properly installed and maintained, should approximate the life of the building. End of service life is based on component condition more than on age.

The scope of the structural inspection is limited. Most inspectors are not engineers or architects. Inspectors who do not possess a valid license to provide these services must avoid using words that could be construed or implied as expressing an opinion about the adequacy of structural components, or about any condition that requires expertise in engineering or architecture (see also Clauses 3.2.B and 13.2.A.3). Inspectors who have valid engineer or architect licenses should not report conclusions or recommendations using those licenses, unless the inspector performs the out-of-scope data gathering and analysis required to make the conclusions or recommendations (see also Clauses 3.2.A, 3.2.B and 13.2.B.2). Most inspectors are not code officials. Only a code official may determine if a condition is a building code violation. Inspectors are not required to, and should not, state or imply that a condition is a code violation (see also Clause 13.2.A.8). Inspections, including structural component inspections, are not technically exhaustive (a defined term) (see also Clause 13.1.B.1). Activities such as measuring components, researching building codes, researching manufacturer's instructions, and performing engineering-type tests and calculations are out-of-scope (see also Clause 13.2.A.8).

Structural Components Not Functioning Properly

Structural components that are not functioning properly present visual evidence of this condition. Such evidence may include significant cracking in foundation footings, walls, and slabs. Such evidence may include significant deformation such as bowing, bulging, rotation, and uplift. Deformation may be defined as a permanent change in the shape of a structural component compared to the original or intended shape. Such evidence may include significant deflection. Deflection may be defined as a temporary change in the shape of a structural component, especially a change that results from a temporary load. Note that building codes allow deflection that may seem significant, but is in fact within allowed limits.

There are no generally accepted definitions of what is considered a significant crack, deformation, or deflection that inspectors can rely upon to identify structural components that are not functioning properly. This is true because analysis of improperly functioning structural components often requires more information than just the presence of deformation, deflection, and cracks. Inspectors should consider erring on the side of reporting structural components that may not be functioning properly because repair of such components can be very expensive.

Significantly Deficient Structural Components

Structural components that are deteriorated or damaged due to causes such as water infiltration, condensation, wood destroying organisms, excessive notching and boring, and physical damage during or after installation are often significantly deficient. Identification and reporting of these components is usually straightforward.

Identification of less obvious significantly deficient structural components presents a challenge for inspectors because many of the criteria that could be used to identify significant deficiencies, such as building codes and manufacturer's installation instructions, are out-of-scope (see also Clauses 13.1.B.1, 13.2.A.3, and 13.2.B.8). Inspectors who have the experience and training to use out-of-scope criteria may do so (see also Clause 2.3.A), but are not required to do so. Inspectors may use professional judgment about identifying and reporting significantly deficient structural components based on out-of-scope criteria; however, such identification and reporting is not required by the American Society of Home Inspectors (ASHI) Standard of Practice for Home Inspections (HISOP).

3.1.B – STRUCTURAL COMPONENT DESCRIPTIONS

This clause requires the inspector to describe the structural components listed in this clause. As previously discussed in Clause 2.2.1.B, a description should allow the reader to distinguish the described system and component from similar systems and components. Foundation descriptions should include the system type and primary materials. The remaining structural descriptions should include the primary materials. Details such as the size of the materials are not required. If multiple systems and materials are used, the inspector should describe each visible system and material. Specifying where the systems or materials are located is not required.

The following descriptions are not intended to be a complete list of all structural systems and materials.

Typical Foundation Descriptions

Typical foundation system descriptions include: slab-on-grade, slab-on-stem-wall, post-tensioned slab, crawlspace, and basement. Typical foundation materials include: concrete, concrete masonry units (aka concrete blocks), insulating concrete forms, brick, stone, preservative-treated wood, and clay tile.

Footings are almost always concealed; therefore, description of footings is not required. Visible absence of a footing is a significant deficiency, and should be reported as such.

Typical Floor Structure Descriptions

Typical floor material descriptions include: dimensional lumber, I-joists, concrete, trusses, and engineered wood products (EWP) that are often used as beams in newer buildings. Describing the type of EWP is not required. Typical floor sheathing materials include: dimensional lumber, plywood, oriented strand board (OSB), and particleboard.

Typical Wall Structure Descriptions

Typical wall material descriptions include: dimensional lumber, concrete masonry units (aka concrete blocks), brick, stone, insulating concrete forms, structural insulated panels, and engineered wood products (EWP) that are often used as beams in newer buildings. Describing the type of EWP is not required. Typical wall sheathing materials include: dimensional lumber, plywood, oriented strand board (OSB), gypsum board, insulating foam sheets, and fiberboard. Describing concealed sheathing is not required. Describing the interior sheathing and wall coverings is not required. The inspector may describe whether the wall structure is platform-framed or balloon-framed, but this is not required.

Typical Ceiling Structure Descriptions

Typical ceiling material descriptions include: dimensional lumber, I-joists, trusses, and engineered wood products (EWP) that are often used as beams in newer buildings. Describing the type of EWP is not required. Ceilings are not usually sheathed. Describing the interior sheathing and wall coverings is not required.

Typical Roof Structure Descriptions

Typical roof material descriptions include: dimensional lumber, I-joists, trusses, and engineered wood products (EWP) that are often used as beams in newer buildings. Describing the type of EWP is not required. Typical roof sheathing materials include: dimensional lumber, plywood, and oriented strand board (OSB).

3.2.A AND 3.2.B – STRUCTURAL ADEQUACY STATEMENTS

As previously discussed, these clauses affirm that inspectors are not required to, and should not, make statements and recommendations about structural component adequacy, or about any condition that requires expertise in engineering or architecture (see also Clauses 13.1.B.1, 13.2.A.3 and 13.2.B.2). This is true even for inspectors with valid licenses that allow them to make such statements and recommendations. Inspectors with valid licenses may do so (see also Clause 2.3.A), but this is not recommended because of the increased risk.

3.2.C – CRAWLSPACE INSPECTION LIMITATIONS

This clause allows inspectors not to enter crawlspace areas with less than twenty-four-inches clearance between the crawlspace floor and an obstruction, or not to enter crawlspaces with an access opening that is less than sixteen-inches by twenty-four-inches (see also Clauses 13.2.A.1, 13.2.D.1, 13.2.D.2, and 13.2.F.1). Many significant deficiencies are found in crawlspaces, so many inspectors enter crawlspaces with lower clearances and smaller access openings. Inspectors may enter crawlspace areas that are not required to be entered. As previously discussed, inspectors must report the crawlspace inspection method, including describing specific areas that were not inspected, and the reason why they were not inspected.

3.2.D – ATTIC INSPECTION LIMITATIONS

This clause allows inspectors not to enter attic areas if doing so requires walking on concealed components (see also Clauses 13.2.A.1, 13.2.D.1, 13.2.D.2, and 13.2.F.1). Many significant deficiencies are found in attics; however, attic inspection methods are less consistent among inspectors. There are also more chances for injury and property damage when traversing concealed attic components. Inspectors have more latitude to determine the attic inspection method compared to crawlspaces. As previously discussed, inspectors must report the attic inspection method, including describing specific areas that were not inspected, and the reason why they were not inspected.

HISOP Section 4 - Exterior

INTRODUCTION

This section contains the requirements and specific limitations for inspecting the building exterior, including components located on the property upon which the building is located. Deficiencies on neighboring buildings and property are out-of-scope. Inspectors may consider reporting obvious and unusual significant deficiencies on neighboring buildings or land that have a high probability of adversely affecting the building or property being inspected, but this is not required.

4.1.A.1 – WALL COVERING, TRIM, AND FLASHING INSPECTION REQUIREMENTS

The objective of the wall coverings, trim, and flashing inspection is to identify and to report visual evidence that these components may not be functioning properly, or that they may be significantly deficient. These components rarely create an unsafe condition, but if one exists, the inspector should report the unsafe condition, usually as an implication of the improper functioning or significant deficiency that is causing the unsafe condition (see also Clause 2.2.1.B.1, unsafe). The service life of these components, if properly installed and maintained, may or may not approximate the life of the building. End of service life is based on component condition more than on age.

Wall covering is a defined term. Wall coverings consist of a relatively thin layer of material that is either affixed to the structural wall system, or is supported on the foundation or on framing. Brick veneer, adhered masonry veneer (aka artificial stone), vinyl siding, and hardboard siding are examples of wall coverings. Wall coverings serve an aesthetic function. They also serve the practical functions of shedding water and protecting the water-resistive barrier installed behind the wall covering.

Some buildings do not have a wall covering. Structural brick, structural stone, buildings that use exposed wood structural panel sheathing, and log homes are examples. In these situations, the structural inspection is also the exterior wall inspection. There is no wall covering to describe.

Paint, stain, and other coatings are not wall coverings. Structural brick is sometimes painted. Log home exteriors are usually stained or sealed with a coating. Describing coatings is not required.

Flashing that is required to be inspected by this section includes wall penetration flashing, such as around windows and doors, and flashing at the intersection of roof coverings and sidewalls. Flashing of roof penetrations, such as plumbing vents, combustion vents, chimneys, and skylights, is addressed in Section 5.

Trim that is required to be inspected by this cause includes brick molding, J-channel, and similar components around windows, doors, and other wall penetrations, and inside and outside corner boards and similar vertical components. Cornice components such as the soffit, fascia, and frieze are covered in Clause 4.1.A.4.

Wall Covering, Trim, and Flashing Not Functioning Properly

The primary functions of wall covering, trim, and flashing are to protect the water-resistive barrier, and to keep water out of the house. It can be difficult to determine if these components are functioning properly because the results of improper functioning are often concealed. Visible water stains, water damage, and fungal growth can be indications that these components are not functioning properly. Further evaluation is often required to determine the cause of water stains, water damage, and fungal growth, so inspectors usually should report the presence of these deficiencies and recommend evaluation to determine the cause.

Significantly Deficient Wall Covering, Trim, and Flashing

Some common examples of significant deficiencies include: damage and deterioration, improper clearance to roof coverings, grade, and hardscape, improper flashing and sealants (caulk) around penetrations and between different types of materials, and cracks in the mortar and in the bricks or stone of brick and stone veneer. Improper installation is a common significant deficiency, although verifying that installation conforms to manufacturer's instructions is out-of-scope (see also Clause 13.2.A.8). Inspectors may consider deterioration or poor application of paint and stain as a significant deficiency if the condition may have a major impact on the wall covering service life.

Wall Covering, Trim, and Flashing End of Service Life

Most wall coverings, trim, and flashing can have a long service life if properly installed and properly maintained. Some wall coverings, such as hardboard siding, and some flashing, such as thin gauge aluminum flashing, can have a service life of less than twenty years if improperly installed and improperly maintained. Even these materials can have a long service life when properly maintained under good conditions.

4.1.A.2 – EXTERIOR DOOR INSPECTION REQUIREMENTS

The objective of the exterior door inspection is to identify and to report visual evidence that these components may not be functioning properly, or that they may be significantly deficient. These components rarely create an unsafe condition, but if one exists, the inspector should report the unsafe condition, usually as an implication of the improper functioning or significant deficiency that is causing the unsafe condition (see also Clause 2.2.1.B.1, unsafe). The service life of these components, if properly installed and maintained, may approximate the life of the building. End of service life is based on component condition more than on age.

Exterior doors include all doors between conditioned and unconditioned spaces. The door from the garage into the building is included, as is the service door between the garage and the exterior, if this door is present. Doors between conditioned space and attics and crawlspaces are included, but hatch covers between these spaces are not included. Hatch covers are addressed in American Society of Home Inspectors (ASHI) Standard of Practice for Home Inspections (HISOP) Section 11. The garage vehicle door is addressed in HISOP Section 10.

Inspection of installed screen doors and storm doors is in-scope. These doors are not required, so absence of these doors is not a deficiency.

Exterior Doors Not Functioning Properly

One primary function of exterior doors is to keep air, water, and vermin out of the house. The door is not functioning properly if light or water stains are seen, or if air flow is felt around the door or through the door.

Another primary function of exterior doors is to provide security, and to provide reasonably easy egress and entrance to the building. Exterior door locks should operate with reasonable ease. Exterior doors should open and close without scraping or rubbing on headers, jambs, or the threshold. Note that this can be a door deficiency, or it can be an indication of a structural deficiency.

Significantly Deficient Exterior Doors

Some common examples of significant deficiencies include: damage and deterioration, and absent and damaged weather stripping and thresholds.

Unsafe Exterior Doors

As previously discussed (see also Clause 2.2.B.1, unsafe), lack of safety glazing in doors is often considered an unsafe condition regardless of the age of the door. Inspectors may consider lack of safety glazing around doors as an unsafe condition based on their professional judgment about the door and the building.

All houses must have an egress door. This door is usually the front door. The egress door may not be equipped with a locking device that requires a key, tool, or special knowledge to operate the device from the interior side. This requirement does not apply to other exterior doors, except for doors that provide emergency egress and entry for bedrooms.

Doors should not open over steps. There are some exceptions to this general rule, such as for screen doors and storm doors serving exterior doors.

4.1.A.3 – DECK, BALCONY, STOOP, STEPS, PORCH, AND RAILING INSPECTION REQUIREMENTS

The objective of inspecting these components is to identify and to report visual evidence that these components may not be functioning properly, or that they may be significantly deficient. These components often create an unsafe condition, and if one exists, inspectors should report the unsafe condition, usually as an implication of the improper functioning or significant deficiency that is causing the unsafe condition (see also Clause 2.2.1.B.1, unsafe). These components can have a limited service life, but end of service life is based on condition more than on age.

It can be difficult to determine if some decks, balconies, stoops, steps, porches and railings are in-scope of a home inspection. Components that rely on the building for part or all of their structural support are in-scope. Components that are supported independently of the building, but are close enough to the building to be considered a functional part of the building are also in-scope. Components such as decks, steps, and railings are sometimes located at a distance from the building. Many, but not all, inspectors consider these components in-scope. Inspectors may decide whether to inspect these components. If the inspector decides not to inspect one of these components, the inspector should report that the component was not inspected, state that the reason is because the component is out-of-scope, and recommend that it be inspected (see also Clause 2.2.B.4).

Exterior step, stairway, guard, and handrail requirements are the same as those for interior components.

Deck, Balcony, Stoop, Step, Porch, and Railings Not Functioning Properly

The primary functions of these components include: supporting design loads, both vertical and horizontal, providing a safe path while traversing these components, and preventing falls. These components are not functioning properly if they are loose, subject to excessive deflection, or do not provide a graspable surface.

Significantly Deficient Deck, Balcony, Stoop, Step, Porch, and Railings

Some common examples of significant deficiencies include: deck ledger attached using only nails, deck ledger improperly attached using bolts or screws, deck ledger without lateral load connectors installed, railing support not connected using tension ties and bolts, stairs not connected using attachment connectors, uneven stair riser height or tread depth, lack of a solid landing at the bottom of a stairway, handrails not graspable, deterioration and rust on fasteners, and stoops uplifted or rotated.

Unsafe Deck, Balcony, Stoop, Step, Porch, and Railings

Many significant deficiencies in these components are also unsafe conditions. When inspectors report a deficiency as both significantly deficient and unsafe, both implications should be explained so that the client may understand the safety implication and the significant deficiency implication.

Deck, Balcony, Stoop, Step, Porch, and Railings End of Service Life

The service life of concrete and masonry components should approximate the life of the building. The service life of wood-constructed components can vary depending on several factors. Open decks can have a service life of ten to twenty years. The service life of covered porches can approximate the life of the building.

Reporting end of service life for these components based on an arbitrary end of service life estimate is not required. If the components are in good condition, reporting end of service life is unnecessary. If the components are in poor condition, inspectors should report the poor condition, and report that the components may be near the end of their service life.

4.1.A.4 – EAVES, SOFFIT, AND FASCIA INSPECTION REQUIREMENTS

The objective of inspecting these components is to identify and to report visual evidence that these components may not be functioning properly, or that they may be significantly deficient. These components rarely create an unsafe condition, but if one exists, the inspector should report the unsafe condition, usually as an implication of the improper functioning or significant deficiency that is causing the unsafe condition (see also Clause 2.2.1.B.1, unsafe). The service life of these components, if properly installed and maintained, should approximate the life of the building. End of service life is based on component condition more than on age.

Inspection of these components is visual from the walking surface immediately below the components. Some inspectors look at these components with binoculars when the components are too high to see clearly from the ground. Some inspectors use a long rod or device to allow probing components that are out-of-reach, but this is not required, nor is accessing these components from a ladder.

Eaves are the extension of rafters beyond the building wall. The soffit is the exposed surface on the underside of the eaves. Fascia (eaves fascia) is a generally vertical trim board attached to the ends of rafters or trusses. The soffit, fascia, frieze, and trim moldings all comprise the cornice. All of these components are in-scope as envisioned by this clause. Eaves and associated trim are not required, and lack of these components is not a defect.

Eaves, Soffit, and Fascia Not Functioning Properly

These components, if present, are primarily decorative. If present, they should be installed and maintained so that animals cannot enter the building through openings in and between components. These components are not functioning properly if there are visible holes through which animals can enter the building.

Significantly Deficient Eaves, Soffit, and Fascia

Some common examples of significant deficiencies include deterioration and damage.

4.1.A.5 – VEGETATION, GRADE, DRAINAGE, RETAINING WALL INSPECTION REQUIREMENTS

The objective of inspecting these components is to identify and to report visual evidence that these components may not be functioning properly, or that they may be significantly deficient. These components rarely create an unsafe condition, but if one exists, the inspector should report the unsafe condition, usually as an implication of the improper functioning or significant deficiency that is causing the unsafe condition (see also Clause 2.2.1.B.1, unsafe). The service life of vegetation is not applicable. The service life of grade and drainage, if properly installed and maintained, should approximate the life of the building. The service life of concrete and masonry retaining walls, if properly installed and maintained, should approximate the life of the building. The service life of wood retaining walls depends on the type of wood used, on installation quality, and on environmental conditions. End of wood retaining wall service life is based on retaining wall condition more than on age.

Inspection of these components is in-scope only if a deficiency or failure would adversely affect the building. Deficiencies or failure of any of these components may adversely affect the building on urban and on smaller suburban lots. Deficiencies or failure of these components is less likely to adversely affect the building if the components are located far from the building on large suburban lots, and on country lots and land. Determining vegetation species and characteristics is out-of-scope.

The inspector should use good professional judgment to determine which deficiencies may adversely affect the building. Many inspectors report obvious, visible deficiencies in these components regardless of their location relative to the building, but this is not required. If the inspector does not inspect a retaining wall that is visible from the building because a deficiency or failure may not adversely affect the building, the inspector should report that the retaining wall was not inspected, and the reason why it was not inspected.

Vegetation Not Functioning Properly

Vegetation is primarily decorative. Not functioning properly does not apply.

Significantly Deficient Vegetation

Some common examples of significant deficiencies include: vegetation that touches or over-hangs the building, or appears likely to do so in the near future, trees or plants located too close to the building, and trees that appear obviously dead.

Grade and Drainage Not Functioning Properly

Proper functioning of grade and drainage is difficult to inspect unless it is raining during the inspection. Evidence of improperly functioning grade and drainage often presents as moisture, or moisture related structural deficiencies, in and around the building, and especially in and around the foundation. Improperly functioning grade and drainage, and improperly functioning gutters and downspouts, are two of the most common causes of building foundation problems, and of damp basements and crawlspaces.

Significantly Deficient Grade and Drainage

Some common examples of significant deficiencies include inadequate provisions for moving rain water away from the foundation, such as inadequate slope away from the building, or lack of other water management systems such as swales, catch basins, and underground drains (French drains).

Retaining Walls Not Functioning Properly

A retaining wall may not be functioning properly if it is not retaining the soil behind the wall. Significant rotation from the top, bulging, slippage off the footing (if present), and undercutting of soil below the wall are typical indications of a retaining wall that is not functioning properly.

Significantly Deficient Retaining Walls

Most retaining wall significant deficiencies fall into two categories. One category is construction defects. The other category is deterioration and damage. Examples of construction defects include failure to install components, such as tiebacks and deadmen, failure to provide lateral load restraint, inadequate footings (often not visible), and use of improper materials, such as tall walls using landscape blocks. Retaining walls built using wood will eventually deteriorate and will require replacement in most environments.

Unsafe Retaining Walls

Retaining walls rarely cause unsafe conditions; however, a guard may be appropriate for a retaining wall with a walking surface on one side and a tall drop on the other side. Whether a guard is required in this situation is subject to local regulations and interpretations.

4.1.A.6 – WALKWAY, PATIO, DRIVEWAY INSPECTION REQUIREMENTS

The objective of inspecting these components is to identify and to report visual evidence that these components may not be functioning properly, or that they may be significantly deficient. These components sometimes create an unsafe condition, and if one exists, the inspector should report the unsafe condition, usually as an implication of the improper functioning or significant deficiency that is causing the unsafe condition (see also Clause 2.2.1.B.1, unsafe). The service life of these components, if properly installed and maintained, may approximate the life of the building. The service life of some types of these components, such as components made from soil or gravel, may be only a few years. End of service life is based on component condition more than on age.

The intent of this clause is that in-scope components are those that are near the building and that provide access to the building. Components located on public easements, such as public sidewalks, and on common areas maintained by neighborhood associations, are out-of-scope (see also Clause 13.2.E.6). Pathways to out-of-scope structures, such as storage sheds and bodies of water, are out-of-scope. The inspector may consider reporting obvious, visible significant deficiencies in out-of-scope components, but this is not required.

Walkway, Patio, Driveway Not Functioning Properly

The primary function of these components is to provide safe and reasonably convenient access to the building. This access should be provided under normal and reasonably foreseeable conditions such as darkness and precipitation. Conditions such as trip hazards, surfaces that may be slick when wet, surfaces that retain water, and severely deteriorated surfaces may be considered as not functioning properly.

Significantly Deficient Walkway, Patio, Driveway

Some common examples of significant deficiencies include: inadequate slope away from the building, walkway, patio, or driveway higher than the grade level between the component and the building, and damage and deterioration. Walkways and driveways made of soil may be significantly deficient, especially if ruts or depressions are present, or if safe access to the building may be difficult during and after precipitation.

Unsafe Walkway, Patio, Driveway

Many functional deficiencies and significant deficiencies in these components are also unsafe conditions. When inspectors report a deficiency as a functional or a significant deficiency, and as an unsafe condition, the inspector should report both the deficiency and the safety implications so that the client understands the implications.

4.1.B – WALL COVERING DESCRIPTIONS

This clause requires the inspector to describe the wall coverings. As previously discussed in Clause 2.2.1.B, a description should allow the reader to distinguish the wall covering from similar wall coverings. Description of wall coverings that are made from the same material, but come in different forms, should include both the material and the form. For example, fiber cement siding comes in panels that are installed vertically, and in planks that are installed horizontally. Details such as the size or thickness of the wall covering are not required. If multiple wall coverings are used, the inspector should describe each wall covering. Specifying where the wall coverings are located is not required.

Exterior insulation finish system (EIFS) looks like stucco, and can be difficult to visually distinguish from one-coat and two-coat stucco systems. Inspectors should attempt to identify EIFS, especially in markets where EIFS is common. If uncertain whether the wall covering is stucco or EIFS, inspectors should report the uncertainty and recommend evaluation to determine the wall covering type and its condition. EIFS inspections are technically exhaustive and require equipment and training beyond that required for a home inspection.

Stucco consists of three different systems, one-coat, two-coat, and three-coat. Visually determining the stucco system is difficult, and is not required.

Typical Wall Covering Descriptions

The following descriptions are examples of wall covering descriptions, and are not intended to be a complete list of all wall coverings. Typical wall covering descriptions include: adhered masonry veneer (also called adhered concrete masonry veneer), aluminum lap siding, brick veneer, exterior insulation finish system (EIFS), fiber cement lap siding, fiber cement panel siding, hardboard lap siding, hardboard panel siding, plywood panel siding, stone veneer, stucco, vinyl lap siding, vinyl panel siding, wood plank siding, and wood shingle siding.

4.2.A – SCREENS, SHUTTERS, AWNINGS, ACCESSORIES INSPECTION LIMITATION

Insect screens are often removed by sellers for various reasons, or were not installed when the building was built. Insect screens are not required on windows and doors. Inspectors may consider reporting significantly damaged or deteriorated installed insect screens on windows and in screened porches, but this is not required. Inspection of installed screen doors and storm doors is in-scope. These doors should be inspected like any other exterior door (see also Clause 4.1.A.2). The absence of a screen door or storm door is not a reportable deficiency, since these doors are not required.

Most shutters are fixed and decorative, and inspection of all decorative components is out-of-scope (see also Clause 13.2.E.3). Inspection of operable shutters, including storm shutters, is out-of-scope. The inspector should report the presence of these shutters, disclaim inspection, and recommend that the seller demonstrate the shutters. The inspector should not operate the shutters because they may be damaged before or during operation.

Awnings, including fixed and motorized models, are out-of-scope. Inspectors may consider reporting significantly damaged or deteriorated installed awnings, but this is not required. The inspector should report the presence of a motorized or a manually operated awning, disclaim inspection, and recommend that the seller demonstrate the awning. Inspectors should not operate the awning because it may be damaged before or during operation.

4.2.B – FENCE, BOUNDARY WALL INSPECTION LIMITATION

Inspection of structures used to delineate property boundaries is out-of-scope. This includes gates serving these structures. Inspectors may consider reporting significantly damaged or deteriorated components, including poorly operating gates, but this is not required. Inspectors may consider reporting unsafe conditions, such as loose bricks or concrete blocks, but this is not required. Inspectors should consider inspecting fences that serve as the access barrier to swimming pools or spas. This may be required in some jurisdictions. If the inspector elects not to inspect these access barriers, the inspector should recommend that a qualified specialist inspect the pool or spa access barriers.

4.2.C AND 4.2.G – SOIL CONDITION AND EROSION CONTROL INSPECTION LIMITATION

Determining soil types and conditions around and under the building is out-of-scope (see also Clauses 13.1.B.1 and 13.2.A.16). This includes, but is not limited to, conditions such as unstable soils, slipping, sliding, and creeping soils and rocks, expansive soils, and sinkholes. Inspectors should consider reporting obvious, visible evidence of these conditions that are present during the inspection, but this is not required.

Inspection of systems and methods to control soil erosion and to stabilize soil is out-of-scope. Erosion is a common issue on steep hills. Examples of erosion control include terracing, use of large rocks called riprap, mulch, gabion (a retaining wall-like cage filled with rocks or sand), and plants. Inspectors should consider reporting obvious, visible evidence of failure of these systems and methods that are present during the inspection, but this is not required.

Retaining walls are a common type of erosion control system, and are sometimes used in conjunction with systems such as terracing. Inspection of retaining walls is in-scope if a deficiency or failure would adversely affect the building (see also Clause 4.1.A.5); otherwise, inspection is out-of-scope.

4.2.D – RECREATIONAL FACILITIES INSPECTION LIMITATION

Recreational facilities is a defined term. Examples of recreational facilities include swimming pools and spas, saunas and steam baths, playground equipment and facilities, basketball goals, tree houses and play houses, and tennis courts and other sports courts and fields. Inspectors should report the presence of these facilities, disclaim inspection, and recommend inspection by a qualified specialist.

4.2.E – OUTBUILDING INSPECTION LIMITATION

Inspection of detached structures intended for storage of motor vehicles, such as garages and carports, is in-scope (see also Clause 13.1.C). Inspection of all other detached structures is out-of-scope (see also Clause 13.2.E.5). Examples of these structures include site-built and prefabricated storage buildings and sheds, barns and animal shelters, airplane hangers, and boat sheds. Inspectors should report the presence of these structures, disclaim inspection, and recommend inspection.

Some detached structures are equipped with systems such as electricity, water, and HVAC. These systems located in out-of-scope detached structures are also out-of-scope, as are the components serving these systems, such as feeder conductors and pipes.

Inspection of habitable detached structures, such as a guest house, is in-scope if the structure is included in the scope of the inspection agreed-upon with the client. If confronted with a habitable detached structure, the inspector should attempt to confirm that the structure is included in the quoted fee. If not included in the quoted fee, the inspector should attempt to renegotiate the fee to include the structure. If this is not possible, the inspector may disclaim inspection and recommend that the structure be inspected.

4.2.F – SEAWALL, BREAKWALL, DOCK INSPECTION LIMITATION

Inspection of structures near the edge of natural bodies of water such as seawalls, breakwalls, docks, and piers is out-of-scope. Inspectors should report the presence of these structures, disclaim inspection, and recommend inspection.

HISOP Section 5 - Roofing

INTRODUCTION

This section contains the requirements and specific limitations for inspecting roof coverings and related components such as gutters and downspouts, roof penetration flashing, and roof penetrations such as skylights, chimneys, plumbing vents, and combustion vents. Inspection of chimneys and vents in this section deals with the exterior of these components, and with their flashing. Roof structural components and chimney structural components are addressed in Section 3. Other chimney and combustion vent deficiencies are addressed in Clause 6.1.A.4 (water heaters), Clause 8.1.B.2 (heating systems), and Clause 12.1.A.3 (fireplaces). Flashing at the intersection of the roof and a sidewall is addressed in Clause 4.1.A.1. Other plumbing vent deficiencies are addressed in Clause 6.1.A.2.

5.1.A.1 AND 5.1A.3 – ROOF COVERING AND FLASHING INSPECTION REQUIREMENTS

The objective of the roof covering and flashing inspection is to identify and to report visual evidence that these components may not be functioning properly, or that they may be significantly deficient. These components rarely create an unsafe condition, but if one exists, the inspector should report the unsafe condition, usually as an implication of the improper functioning or significant deficiency that is causing the unsafe condition (see also Clause 2.2.1.B.1, unsafe). These components have a limited service life, so end of service life reporting is required, when necessary.

Refer to the roof covering inspection method description discussion in Clause 2.2.B.1 for the discussion of roof inspection access requirements, and for reporting how the roof was inspected.

Most roof coverings are steep-slope roof coverings, usually designed for roofs with a 2/12 slope or greater. Most, but not all, steep-slope roof coverings are water-shedding, meaning that the roof covering is not waterproof. The waterproof layer of most steep-slope roof coverings is the underlayment installed under the roof covering.

Some roof coverings are low-slope roof coverings, usually designed for roofs with at least a ¼/1 slope. These roof coverings are waterproof.

Roof flashing inspection is also addressed in Clauses 4.1.A.1 (roof sidewall flashing) and 5.1.A.1 (skylight, chimney, and roof penetration flashing). The inspection requirements are identical no matter which clause is used to inspect these components.

Roof Covering and Flashing Not Functioning Properly

The primary function of roof covering and flashing is to keep water out of the building. It can be difficult to determine if these components are functioning properly, unless it is raining during the inspection, because the results of improper functioning are often concealed. Visible water stains, water damage, and fungal growth can be indications that these components are not functioning properly. Further evaluation is often required to determine the cause of water stains, water damage, and fungal growth, so inspectors should usually report the presence of these deficiencies and recommend evaluation to determine the cause.

Significantly Deficient Roof Covering and Flashing

Inspectors encounter many different roof covering and flashing significant deficiencies. The nature of these deficiencies depends on the roof covering type and materials. Some common examples of significant deficiencies include: deterioration and damage, absent and improperly installed flashing, and penetrations too close to a valley. Improper installation is a common significant deficiency, although verifying that installation conforms to manufacturer's instructions is out-of-scope (see also Clause 13.2.A.8).

Roof Covering and Flashing End of Service Life

The cost to replace roof coverings and flashing is high, so it is understandable that clients are usually concerned about the roof covering condition, and about its remaining service life. Inspectors are not required to, and should not, speculate about the remaining service life of any system or component (see also Clause 13.2.A.2). Inspectors are required to report about systems and components that appear near the end of their service life. Refer to the discussion of Clause 2.2.B.1 for more discussion about end of service life reporting.

The service life of roof coverings and flashing is a function of the material, environment, installation, and maintenance (some roof coverings require periodic maintenance). Roof covering service life varies from a few years for roll roofing to many decades for high quality slate that is properly installed and maintained. The service life of flashing varies from five to fifteen years for inexpensive aluminum flashing and neoprene plumbing vent boots, to decades for high quality lead and copper flashing. There are, however, general guidelines for the service life of common roof covering materials, which the inspector must modify based on the local environment and the visible condition of the roof covering and flashing.

Inspectors should report roof coverings and flashing that appear to the inspector to be near the end of their service lives. It is possible that this end of life condition could be different for the roof coverings and for flashing because some homeowners attempt to save money by replacing the roof coverings but leaving the old flashing in place, and because flashing is sometimes replaced before the roof covering is replaced.

5.1.A.2 – ROOF DRAINAGE SYSTEM INSPECTION REQUIREMENTS

The objective of the roof drainage system inspection is to identify and to report visual evidence that the system may not be functioning properly, or that it may be significantly deficient. These components rarely create an unsafe condition, but if one exists, the inspector should report the unsafe condition, usually as an implication of the improper functioning or significant deficiency that is causing the unsafe condition (see also Clause 2.2.1.B.1, unsafe). The service life of some roof drainage systems, if properly installed and maintained, can approximate the life of the building. Residential roof drainage systems are often not properly installed, and are rarely properly maintained, therefore, end of service life is based on component condition more than on age.

Roof drainage systems is a defined term. Examples include gutters and downspouts used for steep-slope roofs, and scuppers and interior roof drains that are used for low-slope roofs surrounded by parapet walls.

Gutters and other roof drainage systems are not required for steep-slope roofs, and for low-slope roofs that are designed to drain over the edge of the building. Lack of gutters and downspouts in most markets is, however, a reportable deficiency because clients in these markets expect gutters, because lack of gutters can create foundation problems, and because lack of gutters can result in wet soil that can attract wood destroying organisms.

Determining whether the roof drainage is adequate to remove rain water at a sufficient rate is out-of-scope (see also Clause 13.2.A.3). Note that roof drainage systems are not designed to remove all water that falls during extremely heavy rainfall.

Roof Drainage System Not Functioning Properly

The primary function of the roof drainage system is to capture rain water, and to carry it off the roof and away from the building at least as fast as rain falls, during common rain events. It can be difficult to determine if the system is functioning properly unless it is raining during the inspection. Examples of visible indications that rain may be skipping over, or running behind, gutters include deteriorated soffit and fascia, and a line of displaced soil directly under the gutter. Examples of visible indications that a low-slope roof drainage system may not be draining properly include stains and dirt around water collection devices such as roof drains and scuppers. Foundation moisture problems are another example of a possible roof drainage system that is not functioning properly.

Significantly Deficient Roof Drainage System

Inspectors encounter many different roof drainage system significant deficiencies. The nature of these deficiencies depends on the roof drainage system, and on the roof covering. Some common examples of roof drainage system significant deficiencies include: blockage by debris, deterioration and damage, and improper gutter slope toward the downspout, or improper roof slope toward the collection device.

5.1.A.4 – SKYLIGHT, CHIMNEY, ROOF PENETRATION INSPECTION REQUIREMENTS

The objective of the inspection of these components is to identify and to report visual evidence that they may not be functioning properly, or that they may be significantly deficient. These components rarely create an unsafe condition, but if one exists, the inspector should report the unsafe condition, usually as an implication of the improper functioning or significant deficiency that is causing the unsafe condition (see also Clause 2.2.1.B.1, unsafe). The service life of chimneys (especially masonry chimneys), if properly installed and maintained can approximate the life of the building. Skylights, skylight flashing, chimney flashing, and roof penetration flashing have a limited service life. End of service life is based on component condition more than on age.

Skylights and Skylight Flashing Not Functioning Properly

The primary function of skylights is to allow light to enter the building while keeping water and air out of the building. Skylights that are fogged or translucent (unless designed that way) are not functioning properly.

A few skylights are operable and can augment room ventilation. Operable skylights that do not operate are not functioning properly. Inspectors may elect not to operate an operable skylight because it can be difficult to close it. The inspector should report if the operable skylight was not operated, state the reason why, and recommend that the seller demonstrate skylight operation.

The function of skylight flashing is to keep water out of the building. It can be difficult to determine if the skylight is keeping water out of the building unless it is raining during the inspection. Visible water stains, water damage, and fungal growth can be indications that the skylight, or the skylight flashing, is not functioning properly. Water stains and water damage can also be caused by condensation on and around the skylight. Further evaluation is often required to determine the cause of water stains, water damage, and fungal growth, so inspectors usually should report the presence of these deficiencies, and recommend evaluation to determine the cause.

Significantly Deficient Skylights and Skylight Flashing

Inspectors may encounter many different significant deficiencies in skylights. Deterioration and damage are among the most common examples of significant deficiencies in skylights.

Chimneys and Roof Penetrations Not Functioning Properly

Inspection of chimneys and roof penetrations in the American Society of Home Inspectors (ASHI) Standard of Practice for Home Inspections (HISOP) section involves the condition of the visible exterior of the chimneys, roof penetrations, and their flashing. Functional chimney and roof penetration deficiencies are addressed in other HISOP sections such as Section 6 (plumbing), and Section 12 (fireplaces and fuel-burning appliances).

The function of chimney and roof penetration flashing is to keep water out of the building, just like any other flashing. Refer to Significantly Deficient Chimneys and Roof Penetrations, below, for the discussion of functional deficiencies.

Significantly Deficient Chimneys and Roof Penetrations

Inspectors encounter many different chimney, roof penetration, and flashing significant deficiencies. Chimney and roof penetration deterioration, damage, and improperly installed flashing are among the most common examples of these significant deficiencies.

5.1.B.1 – ROOFING MATERIALS DESCRIPTIONS

This clause requires the inspector to describe the roofing materials. As previously discussed in Clause 2.2.B.1, a description should allow the reader to distinguish the roof covering from similar roof coverings. Description of roofing materials means the roof coverings. Description of the flashing materials is not required. Description of the wood species is not required. If multiple roof coverings are used, the inspector should describe each roof covering. Specifying where the roof coverings are located is not required.

Typical Roof Covering Descriptions

The following descriptions are examples of roof covering descriptions, and are not intended to be a complete list of all roof coverings. Typical roof covering descriptions include: built-up roof, clay tiles, concrete tiles, EPDM, fiberglass shingles, lapped-seam metal, modified bitumen, slate, standing-seam metal, wood shakes, wood shingles.

5.1.B.2 – ROOFING MATERIALS INSPECTION METHOD DESCRIPTION

Refer to the roof inspection description section in Clause 2.2.B.1 for the discussion of this requirement.

5.2.A – ANTENNAS INSPECTION LIMITATION

Inspection of all antennas is out-of-scope, regardless of where the antenna is located on the property. This includes inspection of the antenna, mast, cables, and grounding and bonding conductors. This includes television and satellite antennas, amateur and CB radio antennas, and shortwave listening antennas.

Roof penetrations, such as chimneys and plumbing vents, are not intended to bear the load imposed by antennas, or by guy wires used to secure antennas. This also applies to flags, flag poles, and any other component that imposes an unintended load on the roof penetration. Inspectors should report the presence of components that impose an unintended load on roof penetrations, and recommend removal.

Antennas that are attached to roof coverings without visible sealant may leak at the attachment point. Inspectors may wish to report the presence of this condition, if visible, but this is not required.

5.2.B – FLUE INTERIOR INSPECTION LIMITATION

Inspection of the interior of chimneys, flues, and combustion vents is out-of-scope, unless the interior is readily accessible. Refer to the discussion of readily accessible in Clause 2.2.A.

Limited areas of flues and vents are sometimes readily accessible. Limited areas of masonry chimney flues are sometimes readily accessible from the fireplace opening and from the top of the flue. Inspectors should inspect these areas when possible, but inspectors are not required to use ladders or walk on the roof to access the top of a chimney flue. Inspectors are not required to, and in most cases, should not, remove vent caps and other components to gain access to the vent interior.

5.2.C – ACCESSORY INSPECTION LIMITATION

Inspection of all accessories installed on or around the roof is out-of-scope. Examples of such accessories include: flags and flag poles, heat strips, snow and ice guards, and safety railings.

HISOP Section 6 - Plumbing

INTRODUCTION

This section contains the requirements and specific limitations for inspecting plumbing system components including the water distribution pipes and associated valves and fixtures, drain, waste, and vent (DWV) pipes and associated fixtures, water heaters and their vents, fuel storage tanks, fuel distribution pipes, and associated fuel valves, sewage ejectors, and sump pumps.

6.1.A.1 – INTERIOR WATER DISTRIBUTION SYSTEM INSPECTION REQUIREMENTS

The objective of the interior water distribution system inspection is to identify and to report visual evidence that pipes, fixtures, valves, and faucets may not be functioning properly, or that they may be significantly deficient. These components sometimes create an unsafe condition, and if one exists, the inspector should report the unsafe condition, usually as an implication of the improper functioning or significant deficiency that is causing the unsafe condition (see also Clause 2.2.1.B.1, unsafe). These components can have a limited service life, so end of service life reporting is required, when necessary.

A water supply fixture is a device from which water flows. The most common examples are sink faucets, bathtub spouts, shower heads, and hose bibs (also called sill cocks, hose cocks, and spigots). Inspectors are required to operate readily accessible fixtures that a homeowner would normally operate, such as sink faucets. Inspectors are not required to find and operate fixtures, such as yard hydrants and hose bibs, that are located on the property but are not within, or attached to, the buildings being inspected.

Water supply valves include the main water shutoff valve, water heater cold water shutoff valve, fixture stop valves, such as valves under sinks and at toilets, and bathtub and shower valves. Inspectors are required to operate readily accessible valves that a homeowner would normally operate, such as bathtub and shower valves. Inspectors are not required to operate the main water shutoff valve and stop valves (see also Clause 13.2.C.3).

Water distribution pipes include both pipes and tubes. Pipes are rigid, such as galvanized steel. Tubes are usually flexible, such as PEX. Copper water distribution pipes are also considered tubes, although they are usually rigid.

Inspection of the water meter and the building water service pipe is out-of-scope. The water service pipe usually runs from the water meter or well head into the building, where it connects to the water distribution pipes. The building water service pipe is usually underground, thus concealed (see also Clauses 13.1.B.2.a, 13.2.A.1, and 13.2.E.1). The water meter is usually the property of the water utility.

Inspection of water pipes and fixtures located outside of the buildings being inspected is out-of-scope. These pipes and fixtures include those running to out-of-scope buildings, and pipes and fixtures inside out-of-scope buildings, such as storage sheds and barns (see also Clauses 4.2.E, 13.1.C, 13.2.E.1, and 13.2.E.5).

Measuring water flow and water pressure is out-of-scope (see also Clauses 6.2.C and 13.1.B.2.a). Some inspectors measure water pressure, and doing so in areas where water pressure is known to be low or high is prudent, but this is not required. Inspectors are not required to report excess or insufficient water pressure solely on the basis of measuring static water pressure. The maximum water pressure is 80 pounds per square inch (psi). There is no longer a minimum water pressure; however, lack of sufficient water pressure can cause functional flow deficiencies.

Determining water quality is out-of-scope. Inspectors may collect water samples for water quality testing at a laboratory, but this is not required. Collecting water samples requires using specific procedures to help ensure the quality of the samples. Inspectors who wish to collect water samples should contact the testing laboratory for recommended collection procedures (see also Clause 6.2.B.2).

Inspectors are not required to, and often should not, turn on water service to a building, or water supply to a fixture where the water supply has been turned off (see also Clauses 13.2.C.1, 13.2.C.2, and 13.2.C.3). Water is sometime off because of leaks, because fixtures have been removed, or because the building has been winterized. Turning on water can cause significant damage. The inspector should report if water is turned off, and recommend inspection of the plumbing system, or fixture, when water service is restored.

Water Distribution Components Not Functioning Properly

The primary function of water distribution pipes is to deliver an adequate water supply to fixtures without leaking. It can be difficult to determine if pipes are leaking because the leaks are often concealed. Visible water stains, water damage, and fungal growth can be indications that these pipes are leaking. Further evaluation is often required to determine the cause of water stains, water damage, and fungal growth, so inspectors usually should report the presence of these deficiencies, and recommend evaluation to determine the cause.

The primary functions of water distribution valves, fixtures, and faucets are to allow adequate water flow when open, and to shut off water flow completely when off. Inadequate water flow from a fixture or faucet usually indicates that the fixture or faucet is not functioning properly, but it can also indicate other deficiencies. Leaks from a fixture or faucet when the component is off indicates that the component is not functioning properly.

It can be difficult to determine if out-of-scope valves (such as stop valves) are functioning properly because inspectors are not required to, and in most cases should not, operate these valves (see also Clause 13.2.C.3). These valves sometimes seize in position, and can leak or be damaged if operated.

Determining the condition of water distribution components concealed in concrete slabs and in walls is out-of-scope (see also Clauses 13.1.B.2.a and 13.2.A.1). Some inspectors attempt to determine if concealed components are leaking by observing the water meter when all fixtures are off. Performing this service in areas where slab plumbing leaks are known to be common is prudent, but this is not required.

Minimum water flow rate requirements exist for each water supply fixture. The flow rate requirement is measured at the end of pipe, <u>without the fixture installed</u>. Flow rate examples include: four gallons per minute (gpm) at bathtubs and laundry sinks, 0.8 gpm at bathroom vanity sinks, 2.5 gpm at shower heads, and 1.75 gpm at other sinks. Inspectors are not required to measure water flow to determine if the minimum flow rate is present (see also Clauses 6.2.C and 13.1.B.1).

Inspectors are required to visually estimate if reasonable functional flow is present. Inspectors may consider water flow from a fixture that is less than the minimum flow rate indicated above to be inadequate functional flow, and a reportable deficiency.

Significantly Deficient Water Distribution Components

Some common examples of water distribution pipe significant deficiencies include: improper pipe support, including support interval and incompatible support material, lack of insulation when pipes run through unconditioned spaces, such as attics, crawlspaces, and garages, dissimilar metals in contact, such as copper and steel, and corrosion and mineral deposits, especially at fittings.

Some common examples of water distribution valve, fixture, and faucet significant deficiencies include: hot water flowing when cold water is expected (hot on the left, cold on the right) or as indicated by the valve markings, components that move when operated, and damage and deterioration.

Cross-connections between the water distribution system and the drain, waste, and vent system (DWV) are required to be equipped with a means to prevent backflow from the DWV system into the water distribution system. Some common examples of cross-connections include: tubes installed on faucets that are below the top of the sink or bathtub, bathtub spouts below the top of the bathtub, and absent or improperly installed backflow prevention devices on a landscape irrigation system water supply. Inspectors are required to report cross-connections that do not have adequate backflow protection, such as an air gap or a backflow protection device. Inspectors are not required to test backflow protection devices (see also Clause 13.1.B.1).

Inspectors are not required to identify or to report systems or components that may be subject to recall or controversy (see also Clause 13.2.A.17). Some water distribution pipes, however, are known to be problematic, so many inspectors report the presence of these pipes, but this is not required. Examples of these problematic pipes include polybutylene, especially those installed in the 1980s, and PEX, especially those installed in the early to late 2000s.

Unsafe Water Distribution Components

This is not intended to be a complete list of all potentially unsafe water distribution components.

Bathtub and shower valves are now required to have an internal mechanism, or an external mechanism such as a mixing valve, that limits water temperature to 120° F. or less. This is a relatively recent requirement, so valves in older buildings often do not comply. Inspectors may consider reporting older valves as unsafe, and should consider reporting newer valves as unsafe if the hot water temperature appears excessive. Inspectors are not required to measure water temperature (see also Clause 13.1.B.1).

Water distribution components made from lead are considered unsafe because of the risk of lead poisoning. Inspectors should report the presence of these components and recommend replacement. This does not apply to lead drainage system components.

Water Distribution Components End of Service Life

Mechanical water distribution components such as valves and faucets have a limited service life, but they can often be repaired at a relatively modest cost.

Some water distribution components are known to have a limited service life. Galvanized steel pipes are the most common example. These pipes are known to rust and become clogged with debris after about fifty years. Inspectors should consider reporting these pipes if they appear to be more than fifty years old, which many are. Copper water pipes are known to deteriorate under certain conditions such as aggressive water. Inspectors in markets where this is known to occur may consider reporting these pipes as near end of service life, depending on their estimated age, and on the conditions in which they are installed. Copper water pipe deterioration is more common when the pipes are encased in concrete.

Reporting end of service life for most water distribution components based on an arbitrary end of service life estimate is not required. If the components appear in good condition, reporting end of service life is unnecessary. If the components appear in poor condition, inspectors should report the poor condition, and report that the system may be near the end of its service life.

6.1.A.2 – DRAIN, WASTE, AND VENT (DWV) SYSTEM INSPECTION REQUIREMENTS

The objective of the interior DWV system inspection is to identify and to report visual evidence that pipes and fixtures may not be functioning properly, or that they may be significantly deficient. These components rarely create an unsafe condition, but if one exists, the inspector should report the unsafe condition, usually as an implication of the improper functioning or significant deficiency that is causing the unsafe condition (see also Clause 2.2.1.B.1, unsafe). These components can have a limited service life, so end of service life reporting is required, when necessary.

A DWV fixture is a device that receives water that flows from the water distribution system. The most common examples are sinks, bathtubs, shower pans (also called receptors), and toilets (also called commodes and water closets). Inspectors are required to run water into readily accessible fixtures that a homeowner would normally use.

Inspection of the building sewer pipe is out-of-scope. This pipe usually runs from the municipal sewer or private sewage disposal system (usually called the septic system) into the building, where it connects to the building drain pipe. The building sewer pipe is usually underground, thus concealed (see also Clauses 13.1.B.2.a, 13.2.A.1, and 13.2.E.1).

Inspection of DWV pipes and fixtures located outside of the buildings being inspected is out-of-scope. This includes pipes running to, and pipes and fixtures inside out-of-scope buildings such as storage sheds and barns (see also Clauses 4.2.E, 13.1.C, 13.2.E.1, and 13.2.E.5).

DWV Components Not Functioning Properly

The primary function of drain and waste pipes and fixtures is to remove water from fixtures, and to direct the water into a sewer or septic system without leaking and without blockages. Inspectors are not required to fill fixtures, such as bathtubs and shower pans, to test for leaks, or to test overflow components (see also Clause 6.2.D). It can be difficult to determine if pipes are leaking because the leaks are often concealed. Visible water stains, water damage, and fungal growth can be indications that these pipes are leaking. Further evaluation is often required to determine the cause of water stains, water damage, and fungal growth, so inspectors usually should report the presence of these deficiencies, and recommend evaluation to determine the cause. It is not practical to determine if pipe blockages occur unless a blockage occurs during the inspection.

The primary function of a toilet, a special type of fixture, is to remove a reasonable quantity of solids from the bowl in one flush, without leaking either at the water supply or at the drainage outlet. Leaks at the water supply are usually easy to detect. Leaks at the drainage outlet can be difficult to detect because the leaks are often concealed. Inspectors are required to flush every operating toilet, and to make a visual determination if the flushing action appears normal. Inspectors are not required to measure flushing action, or to determine the force of the flushing action.

The primary function of the vent system is to protect the water seal in traps. The trap water seal keeps noxious and potentially explosive sewer gas out of the building. It can be difficult to determine if the vent system is installed and functioning properly because the vent pipes and vent connections are often concealed.

Determining the condition of DWV components concealed in concrete slabs and in walls is out-of-scope, and is not required (see also Clauses 13.1.B.2.a and 13.2.A.1). There are methods of detecting these leaks, such as camera inspections.

Water removal from fixtures should occur at approximately the same rate or at a greater rate than the rate of water flow into the fixture. This is referred to as functional drainage. Inspectors are required to visually determine if reasonable functional drainage is present. Inspectors may consider water backing up into a fixture such that the fixture is likely to overflow as inadequate functional drainage, and a reportable deficiency.

Significantly Deficient DWV Components

Some common examples of DWV pipe significant deficiencies include: improper pipe support, including support interval and incompatible support material, improper installation of pipes and fittings including improper type of pipe and fitting, improper installation of traps and fixture drains, including improper trap type (such as S traps), dissimilar metals in contact, such as copper and steel, and corrosion and mineral deposits, especially at fittings.

DWV Components End of Service Life

Some DWV pipes are known to have a limited service life. Galvanized steel pipes and cast-iron pipes are the most common examples. These pipes are known to rust, become clogged by debris, and leak after about fifty years. Inspectors should consider reporting these pipes if they appear to be more than fifty years old, which most are.

Reporting end of service life for most DWV components based on an arbitrary end of service life estimate is not required. If the components appear in good condition, reporting end of service life is unnecessary. If the components appear in poor condition, inspectors should report the poor condition, and report that the system may be near the end of its service life.

6.1.A.3 – WATER HEATING SYSTEM INSPECTION REQUIREMENTS

The objective of the water heating system inspection is to identify and to report visual evidence that water heating equipment and hot water distribution system components may not be functioning properly, or that they may be significantly deficient. These components can create an unsafe condition, and if one exists, the inspector should report the unsafe condition, usually as an implication of the improper functioning or significant deficiency that is causing the unsafe condition (see also Clause 2.2.1.B.1, unsafe). System components can have a limited service life, so end of service life reporting is required, when necessary.

Water heating equipment types that are in-scope include storage-tank, demand (tankless), indirect-fired types such as tankless coils installed in hot water boilers, and point-of-use equipment, such as instant hot water faucets. Water heaters that use alternative energy sources such as solar water heaters, including solar assist systems, and ground-source heat pumps are out-of-scope (see also Clause 6.2.A.5).

Hot water distribution pipe inspection is addressed in Clause 6.1.A.1. Hot water circulation systems are in-scope, including pumps, valves, and pipes. These systems often use the water distribution pipes, although separate pipe loops are sometimes installed in new construction.

Inspection of water heating systems located outside of the buildings being inspected is out-of-scope (see also Clauses 4.2.E, 13.1.C, and 13.2.E.5).

Water Heating System Not Functioning Properly

The primary function of the water heating system is to safely produce an adequate supply of hot water, and to deliver that water to the points of use at a safe temperature. The maximum safe domestic hot water temperature where people come in direct contact with the water (at faucets) is generally considered to be 120° F, although there are exceptions. Higher water temperature is recommended for appliances such as dishwashers. Excessive water temperature at faucets is considered a reportable deficiency, although the deficiency may be in the faucet or the valve, not the water heating system.

There is no generally accepted minimum domestic hot water temperature, although many would consider a temperature below 100° F at the point of use to be deficient. It is out-of-scope for home inspectors to determine if the hot water supply is adequate, and if the time required for hot water to reach a fixture is acceptable. Home inspectors are not required to measure water temperature (see also Clauses 13.1.B.1 and 13.2.A.3).

Significantly Deficient Water Heating System

Some common examples of water heating system significant deficiencies include: various defects in the installation of the temperature/pressure relief valve and its extension pipe, tank leaks, rust, damage and deterioration, and soot inside or outside of the cabinet, lack of a thermal expansion device when required, lack of vehicle impact protection, poor support, and inadequate combustion air.

Unsafe Water Heating System

The primary water heater safety device is the temperature/pressure relief valve and its associated extension pipe. Inspection of these components is visual. Inspectors are not required to, and in most cases should not, test the valve (see also Clause 13.2.C.4). Water from the valve can damage nearby components, and sometimes the valve will not completely close after testing (although this is a significant deficiency). Note that testing the valve may be required in some jurisdictions.

Other water heater safety-related deficiencies are addressed in other HISOP sections such as Clause 6.1.A.4, and Clause 6.1.A.5. High water temperature at faucets is sometimes a faucet/valve deficiency (e. g., no mixing valve).

Water Heating System End of Service Life

While the cost to replace a water heater is not usually significant, clients are, nonetheless, usually concerned about the water heater, and about its remaining service life. Inspectors are not required to, and should not, speculate about the remaining service life of any system or component (see also Clause 13.2.A.2). Inspectors are required to report about systems and components that appear to be near to be the end of their service life. Refer to Clause 2.2.B.1 for more discussion about end of service life reporting.

The service life of a water heater is a function of the water heater type, environment, water quality, maintenance, and installation. The average service life of storage-tank type domestic water heaters is around fifteen years; however, actual service life varies widely. Inspectors should report water heaters that appear to be near the end of service life based on water heater age (if determined) and on physical condition. Note that determining water heater age is not required, and is sometimes not possible.

6.1.A.4 – VENT SYSTEM INSPECTION REQUIREMENTS

The objective of the vent system inspection is to identify and to report visual evidence that the vent system may not be functioning properly, or that it may be significantly deficient. This includes all appliance vent systems such as systems serving water heaters, furnaces, and boilers. These components can create an unsafe condition, and if one exists, the inspector should report the unsafe condition, usually as an implication of the improper functioning or significant deficiency that is causing the unsafe condition (see also Clause 2.2.1.B.1, unsafe). The service life of these components, if properly installed and maintained, should approximate the life of the building. End of service life is based on component condition more than on age.

Vent systems that rely on hot exhaust gas temperatures (natural draft) to operate usually consist of a vent connector and a vent. A draft hood may be part of the system. The vent is the last vertical component in the vent system. The component between the appliance and the vent is the vent connector. The vent may be a manufactured component, such as a double-wall vent, or it may be a chimney. Vent connectors are usually single-wall or double-wall metal pipes. Inspection of the readily accessible parts of all of these components is in-scope.

Prescriptive rules (code) exist governing vent system installation for natural draft vent systems (Category I gas water heaters, and oil-fired water heaters). Inspectors are not required to determine compliance with codes and regulations (see also Clause 13.2.A.8). Codes and regulations can, however, be a source of information for identifying significant deficiencies.

Vent system installation and termination for appliances, such as many demand (tankless) water heaters and high efficiency furnaces and boilers, is addressed exclusively by manufacturer's instructions. Installation and termination of vent systems that rely on positive pressure in the vent system, and the installation and termination of vent systems that rely on a fan to draw exhaust gasses through the vent system (induced draft system) are also addressed exclusively by manufacturer's instructions. These instructions vary between manufacturers and between models from the same manufacturer. Inspectors are not required to determine compliance with manufacturer's instructions (see also Clause 13.2.A.8). Inspectors should inspect for visual evidence that these vent systems are not functioning properly, but confirming installation and termination of these systems per manufacturer's instructions is out-of-scope.

Vent System Not Functioning Properly

The primary function of most vent systems is to safely expel combustion gasses out of the building without allowing the gasses to cool to the point where they condense inside the vent, or where the gasses fall down the vent and back into the building or into the appliance (backdrafting). Backdrafting can sometimes be detected during the inspection when condensation is observed on a mirror placed near the draft hood. Soot, debris, rust, and stains on and around vent system components can indicate vent system functional deficiencies. Note that condensation is expected in some high-efficiency appliance vent systems.

Significantly Deficient Vent System

Some common examples of vent system significant deficiencies include: disconnected components, components not properly secured to each other, components too close to combustible materials or insulation, vent termination too close to obstruction, damage or deterioration, vent connector does not slope up toward vent, improper use of single-wall vent material, improper vent connector connection to chimney, and improper use of chimney as a vent.

6.1.A.5 – FUEL STORAGE AND DISTRIBUTION SYSTEM INSPECTION REQUIREMENTS

The objective of the fuel storage and distribution system inspection is to identify and to report visual evidence that this system may not be functioning properly, or that it may be significantly deficient. These components can create an unsafe condition, and if one exists, the inspector should report the unsafe condition, usually as an implication of the improper functioning or significant deficiency that is causing the unsafe condition (see also Clause 2.2.1.B.1, unsafe). System components can have a limited service life, so end of service life reporting is required, when necessary.

In-scope systems include those for gas (natural gas and propane), and those for oil. Aboveground storage tanks for oil are in-scope. Visible fuel distribution components, such as pipes and valves, for both oil and gas are in-scope. Underground tanks for oil are out-of-scope. Fuel distribution components located underground, and those that are concealed, for both oil and gas, are out-of-scope (see also Clause 13.1.B.2.a, 13.2.A.1, 13.2.A.8, and 13.2.E.1).

Inspectors are not required to determine the presence of underground fuel tanks (see also Clause 13.2.E.1); however, most inspectors in markets where oil and propane are common report visible indications of underground tanks, such as fill pipes and vent pipes. Visually inspecting for indications of underground fuel tanks, while not required, is prudent.

Components located before the point of fuel delivery are out-of-scope. The point of fuel delivery in a utility-supplied natural gas system is on the output side of the gas meter. The gas meter, regulator, and supply pipe are the property of, and are under the control of, the utility. The point of fuel delivery in a propane system is where the fuel pipe enters the house. The propane tank is often the property of, and under the control of, the propane supplier.

Most inspectors observe the gas meter and the regulator, and report obvious deficiencies, including gas leaks detected by smelling gas. Most inspectors observe the propane tank and the regulator, and report obvious deficiencies, including gas leaks, detected by smelling gas. None of these observations are required.

Use of devices to test for gas leaks is out-of-scope (see also Clause 13.1.B.1). Some inspectors use devices to test for gas leaks, but this is not required.

Fuel Storage and Distribution System Not Functioning Properly

The primary function of the fuel storage and distribution system is to safely store and deliver an adequate amount of fuel to fuel-burning appliances without leaking. It can be difficult to determine if pipes are leaking because the leaks are often concealed. Odors and visible stains can be indications that these pipes are leaking. Further evaluation is often required to determine the cause or source of odors and stains, so inspectors should usually report the presence of these deficiencies, and recommend evaluation to determine the cause.

It is out-of-scope for home inspectors to determine if the fuel storage capacity and fuel distribution pipe sizes are adequate to serve the installed appliances (see also Clauses 13.1.B.1, 13.2.A.3, and 13.2.A.8).

Significantly Deficient Fuel Storage and Distribution System

Some common examples of fuel storage and distribution system significant deficiencies include: damage and deterioration, improper installation of pipes including improper support intervals and improper support materials, improper installation of flexible fuel connectors, such as running connectors through partitions and through appliance cabinets, absent or improper location of fuel valve, and copper gas tubing not labeled as containing gas.

Fuel Storage and Distribution System End of Service Life

The service life of fuel distribution pipes usually approximates the life of the building if properly installed and maintained. The service life of fuel storage tanks depends on the type of tank and its location. Oil storage tanks located inside can last thirty years or more. Oil storage tanks located outside, including buried tanks, can last as little as ten years, with twenty years as a reasonable average. Propane tanks are always located outside. The service life of above ground propane tanks is usually thirty to forty years.

Reporting end of service life for most fuel storage and distribution system components based on an arbitrary end of service life estimate is not required. If the components appear in good condition, reporting end of service life is unnecessary. If the components appear in poor condition, inspectors should report the poor condition, and report that the system may be near the end of its service life.

6.1.A.6 – SEWAGE EJECTOR AND SUMP PUMP INSPECTION REQUIREMENTS

The objective of the sewage ejector and sump pump inspection is to identify and to report visual evidence that these components may not be functioning properly, or that they may be significantly deficient. These components rarely create an unsafe condition, but if one exists, the inspector should report the unsafe condition, usually as an implication of the improper functioning or significant deficiency that is causing the unsafe condition (see also Clause 2.2.1.B.1, unsafe). System components can have a limited service life; however, end of service life is based on component condition more than on age.

Sewage ejectors and sump pumps are sometimes improperly identified. A sewage ejector should be installed when plumbing fixtures are located below the building sewer pipe. The sewage ejector pumps the material up to the building drain. A sump collects ground water in the sump pit, and the sump pump ejects the water out of the building.

Inspectors should operate these systems, if possible and safe to do so. Inspectors should operate sewage ejectors by running water in fixtures that drain to the sewage ejector until they hear the pump activate. Inspectors should operate sump pumps by pressing on the float with an insulated tool such as a screwdriver, if the inspector believes that it is safe to do so. Removing a sewage ejector cover to inspect or test is not required, and is not recommended. Removing a sump pump cover to inspect or to operate is required if the cover qualifies as a readily openable access panel (see also Clauses 13.1.B.2.a, and 13.2.A.1). Operating a sump pump is not required, or recommended, unless the inspector considers it safe to do so (see also Clause 13.2.F.1).

Sewage Ejectors and Sump Pumps Not Functioning Properly

The primary function of sewage ejectors and sump pumps is to eject water and other material to an approved disposal location. This should occur at a rate that is adequate to keep the container from overflowing.

The approved disposal location for a sewage ejector is usually the building drain, or a branch drain. The approved disposal location for a sump pump is outside of the building and away from the foundation. A sump pump should not discharge into the public sewer, unless approved by the local sewer authority, and should not discharge into a septic system.

It is out-of-scope for home inspectors to determine if the sewage ejector system or sump pump system is an adequate size (see also Clauses 13.2.A.3, and 13.2.A.8). It is out-of-scope to test battery backup and alarm systems (see also Clause 13.2.C.4).

Significantly Deficient Sewage Ejectors and Sump Pumps

Some common examples of sewage ejector and sump pump significant deficiencies include: damage and deterioration, improper discharge pipe type, improper discharge pipe termination location, sewage ejector full-open valve and check valve not installed, improper sewage ejector venting, and plumbing fixtures discharging into sump pump.

6.1.B.1, 6.1.B.2, 6.1.B.3 - PLUMBING SYSTEM DESCRIPTIONS

These clauses require the inspector to describe: the interior water distribution pipe materials, the interior drain, waste, and vent pipe materials, the types of water heating equipment and the energy sources, and the location of the main water shutoff valve and the main fuel system shutoff valve. As previously discussed in Clause 2.2.B.1, a description should allow the reader to distinguish the described materials, equipment, and locations from similar materials, equipment, and locations. If multiple types of materials and equipment are used, the inspector should describe each type of material and equipment. Reporting where the materials and equipment are located, except for the shutoff valves, is not required.

Typical Water Distribution Materials Descriptions

The following descriptions are examples of water distribution materials descriptions, and are not intended to be a complete list of all such materials. Typical water distribution materials descriptions include: copper tubing, CPVC pipe, galvanized steel pipe, lead pipe, PEX tubing, polybutylene tubing.

Typical Drain, Waste, and Vent Materials Descriptions

The following descriptions are examples of drain, waste, and vent materials descriptions, and are not intended to be a complete list of all such materials. Typical water distribution materials descriptions include: ABS pipe, cast iron pipe, copper pipe, galvanized steel pipe, lead pipe, PVC pipe.

Typical Water Heating Equipment and Energy Source Descriptions

Describing the capacity (size) of the water heater, or the type water heater, is not required. Describing the location of water heating equipment is not required. Describing the model number and serial number is not required. Describing the type of fuel-gas is not required; however, inspectors may consider describing the fuel-gas type if the fuel is propane, especially if there is no visible propane tank.

The following descriptions are examples of water heating equipment and energy source descriptions, and are not intended to be a complete list of all such descriptions. Typical water heating equipment and energy source descriptions include the following. Hot water is provided by a gas-fired storage tank water heater. Hot water is provided by an electric water heater. Hot water is provided by a demand (tankless) water heater. Hot water is provided by a tankless coil inside the hot water boiler.

Typical Valve Location Descriptions

Describing the precise location of the water and fuel shutoff valves is not required; a general location is acceptable. Directions such as right, left, front, and back are acceptable if the reference point is defined in the report, such as at the street looking at the front of the house. Sometimes it is not possible to locate these valves because they are concealed behind occupant belongings, located in difficult to access places, or for other reasons. Inspectors should report if these valves were not located and recommend that the seller provide their location.

The following descriptions are examples of shutoff valve descriptions, and are not intended to be a complete list of all such descriptions. The main gas shutoff valve is located at the meter in the right side yard. The main fuel oil shutoff valve is located at the oil tank in the basement. The main water shutoff valve is located in the kitchen pantry. The main water shutoff valve is located in the right front area of the basement.

Some inspectors place a tag or label on these valves. This is a good service, but this is not required.

6.2.A.1 – CLOTHES WASHING MACHINE PLUMBING CONNECTION INSPECTION LIMITATION

Inspection of the clothes washing machine water supply and drainage connections is out-of-scope, insofar as testing and operating these connections is concerned. Inspectors are not required to, and should not, disconnect clothes washing machine hoses, operate the clothes washing machine water supply valves, and run water into washing machine drain pipes (see also Clause 13.2.F.1). Visual inspection of the readily accessible clothes washing machine connection components is in-scope as part of the inspection of the water distribution system, and the drain, waste, and vent system.

6.2.A.2 – FLUE INTERIOR INSPECTION LIMITATION

Refer to Clause 5.2.B for the discussion of this limitation.

6.2.A.3, 6.2.A.7, AND 6.2.B.1 – PRIVATE WATER SUPPLY AND SEWAGE DISPOSAL INSPECTION LIMITATION

Inspection of private water supply systems is out-of-scope. This limitation includes the well and well head, well pump, water service pipe from the well to the building, pressure tank, pressure switch, and all other well-related components. This limitation includes determining well capacity, pressure, and flow rate, such as well draw-down tests (see also Clauses 6.2.C, 13.1.B.1, and 13.2.A.3). This limitation includes determining well water quality.

Inspection of private sewage disposal systems is out-of-scope. These systems are usually called septic systems. This includes the septic tank, the disposal field, battery backup systems, overflow alarms, and all other related components. A separate license is required to provide this service in some jurisdictions.

It is also out-of-scope to determine and to report whether a building is served by a public or private water supply system, and whether the building is served by a public or private waste disposal system. If the inspector elects to report about these matters, the report should state that any statement is a belief based on available information, and that the client should have a qualified specialist confirm the statement.

6.2.A.4 – WATER CONDITIONING SYSTEM INSPECTION LIMITATION

Inspection of water conditioning systems is out-of-scope. This includes whole-house and point-of-use filters, water softeners, water conditioners, chemical injectors, and sanitizing systems such as ultraviolet lights. Inspectors should consider reporting the presence of these systems, if visible, disclaiming inspection, and recommending specialist evaluation if the client wants assurance about the installation, condition, and functionality of the system.

6.2.A.5 – ALTERNATIVE ENERGY EQUIPMENT INSPECTION LIMITATION

Inspection of heating and cooling equipment (appliances) that use an energy source other than fuel-gas, oil, or electricity is out-of-scope. This limitation applies to all heating and cooling appliances. This limitation includes solar heating systems, including solar systems that preheat water before delivery to a conventional water heater. This limitation includes ground-source and water-source heat pumps. Inspectors should consider reporting the presence of these appliances, if known, disclaiming inspection, and recommending specialist evaluation if the client wants assurance about the installation, condition, and functionality of these appliances.

6.2.A.6 – FIRE EXTINGUISHING SYSTEM AND LANDSCAPE IRRIGATION SYSTEM INSPECTION LIMITATION

Inspection of fire extinguishing (fire sprinkler) systems is out-of-scope. This includes all system components such as controls, pipes, valves, and heads.

Inspection of landscape irrigation (sprinkler) systems is out-of-scope. This includes all system components such as controls, pipes, valves, and heads. Inspection of the backflow prevention device is in-scope as part of the water distribution system inspection.

It is also out-of-scope to determine if these systems are properly designed, properly installed, and adequate to perform their intended function (see also Clauses 13.1.B.1 and 13.2.A.3)

Inspectors should consider reporting the presence of these systems, if visible, disclaiming inspection, and recommending specialist evaluation if the client wants assurance about the installation, condition, and functionality of these systems.

6.2.B.2 – WATER QUALITY INSPECTION LIMITATION

Determining water quality is out-of-scope. This includes all types of water quality-related issues such as sediment, water chemistry, contaminates, taste, and smell.

6.2.B.3 – COMBUSTION AIR ADEQUACY INSPECTION LIMITATION

Inspection of combustion air components includes visually determining if such components exist, and visually determining if evidence of possible inadequate combustion air is present. Measuring combustion air ducts and openings, and calculating the size of combustion air ducts and openings is out-of-scope (see also Clauses 13.1.B.1 and 13.2.A.3).

6.2.C – WATER MEASUREMENT INSPECTION LIMITATION

Measuring water pressure and flow rate is out-of-scope. Measuring well capacity and flow rate is out-of-scope. Visually estimating functional flow from water supply fixtures is in-scope as part of inspecting the water distribution system.

6.2.D – FIXTURE FILL INSPECTION LIMITATION

Filling fixtures and receptors to test for leaks, and for functional drainage, is out-of-scope. This includes testing sink and bathtub overflow components.

HISOP Section 7 - Electrical

INTRODUCTION

This section contains the requirements and specific limitations for inspecting electrical system components including: building electrical service and grounding components, electrical panelboards and their enclosures, service entrance conductors, feeder conductors, branch circuit conductors, overcurrent devices, lights, switches, and receptacles, ground-fault circuit interrupters, and arc-fault circuit interrupters.

7.1.A.1 – SERVICE DROP INSPECTION REQUIREMENTS

The objective of the electrical service drop inspection is to identify and to report visual evidence that the service drop may not be functioning properly, or that it may be significantly deficient. These components can create an unsafe condition, and if one exists, the inspector should report the unsafe condition, usually as an implication of the improper functioning or significant deficiency that is causing the unsafe condition (see also Clause 2.2.1.B.1, unsafe). The service life of these components, if properly installed and maintained, should approximate the life of the building. End of service life is based on component condition more than on age.

A service drop runs overhead from a utility transformer on a pole to the building. A service lateral runs underground from a utility transformer on the ground to the building. This clause also includes the visible part of the service lateral; however, almost all of a service lateral is underground, thus concealed. Only the visible conduit (the riser) is in-scope (see also Clauses 13.1.B.2.a, 13.2.A.1, and 13.2.E.1). The term service drop includes the service lateral for purposes of this discussion.

Inspectors are not required to determine if the service conductors are the correct size. Service conductor size is determined by the electric utility, not by the National Electrical Code. Service conductors may be smaller than other conductors because service conductors are installed where they can dissipate heat more effectively than other conductors. Inspectors may consider reporting possibly inadequate service conductor size, but should exercise caution if doing so. This situation is most likely to occur at older buildings when the building electrical service is upgraded from thirty-amp or sixty-amp service.

Service Drop Not Functioning Properly

The primary function of service drop conductors is to safely and reliably conduct electricity to the building. If the building is supplied with electricity, then the service drop conductors are usually functioning properly. It is possible for the building to have electricity with an improperly functioning service drop, but this situation is very uncommon in residential buildings.

Significantly Deficient Service Drop

Some common examples of service drop significant deficiencies include: inadequate clearance to roof, to ground, or to operable windows, uninsulated connection to the service entrance conductors, service drop run through or below tree branches, disconnected neutral conductor, and insulation damage or deterioration.

Unsafe Service Drop

Inspectors may consider individual service drop conductors that are subject to accidental contact to be unsafe. This includes service drop conductors that might be reached from the ground, from an open window, and from a deck or balcony. Service drop conductors with exposed energized (uninsulated) conductors are unsafe regardless of their location.

7.1.A.2 – SERVICE ENTRANCE COMPONENT INSPECTION REQUIREMENTS

The objective of the electrical service entrance component inspection is to identify and to report visual evidence that service entrance components may not be functioning properly, or that they may be significantly deficient. These components can create an unsafe condition, and if one exists, the inspector should report the unsafe condition, usually as an implication of the improper functioning or significant deficiency that is causing the unsafe condition (see also Clause 2.2.1.B.1, unsafe). The service life of service entrance components depends on the type of components used, and on environmental conditions. End of service life is based on component condition more than on age.

Service entrance components for a service drop begin near the building at a mast or other point of connection to the building. Service entrance components for a service lateral usually begin at the electric meter.

Service entrance components include the service entrance conductors, and associated cables, masts, weather heads, and raceways.

Service Entrance Components Not Functioning Properly

The primary function of service entrance components is to safely and reliably conduct electricity from the service point to the service equipment (main electricity shutoff). If the building is supplied with electricity, then the service entrance components are usually functioning properly. It is possible for the building to have electricity with improperly functioning service entrance components, but this situation is very uncommon in residential buildings.

Significantly Deficient Service Entrance Components

Some common examples of service entrance component significant deficiencies include: deteriorated or damaged service mast, conductor insulation, weather head, or flashing, loose or damaged meter base, disconnected neutral conductor.

Unsafe Service Entrance Components

Inspectors may consider individual service entrance conductors that are subject to accidental contact to be unsafe. This includes service entrance conductors that might be reached from the ground, from an open window, and from a deck or balcony. This does not include service entrance conductors enclosed in sheathing (SE cable) or in conduit. Service entrance conductors with exposed energized (uninsulated) conductors are unsafe regardless of their location.

7.1.A.4 – SERVICE GROUNDING INSPECTION REQUIREMENTS

The objective of the service grounding system inspection is to identify and to report visual evidence that service grounding components may not be functioning properly, or that they may be significantly deficient. These components rarely create an unsafe condition, but if one exists, the inspector should report the unsafe condition, usually as an implication of the improper functioning or significant deficiency that is causing the unsafe condition (see also Clause 2.2.1.B.1, unsafe). The service life of these components, if properly installed and maintained, should approximate the life of the building. End of service life is based on component condition more than on age.

Service grounding system components are the grounding electrode and the grounding electrode conductor, together referred to as the grounding electrode system. Typical grounding electrodes include metal underground water service pipe, rods and pipes, and a reinforcing bar or a wire encased in the concrete footing (a Ufer ground). Typical grounding electrode conductor is an uninsulated copper wire.

The grounding electrode conductor connection may occur at any accessible location upstream from the service equipment (main disconnect). A grounding electrode conductor connection downstream from the service equipment is usually a significant deficiency.

Much of the service grounding system is often concealed, including connection of the grounding electrode conductor to the grounding electrode. Inspectors are not required to inspect concealed components, or to verify the condition of the entire grounding electrode system (see also Clauses 13.1.B.1, 13.1.B.2.a, 13.2.A.1).

Service Grounding Not Functioning Properly

The primary functions of the service grounding system are to serve as an alternate return path for electricity in case the utility grounded (neutral) conductor fails to function properly, and to help reduce voltage and current anomalies caused by events such as lightning and issues in the electricity transmission grid. It is not possible to determine if the service grounding system is functioning properly during a home inspection, so determining this is out-of-scope (see also Clause 13.1.B.1).

Significantly Deficient Service Grounding

Some common examples of service grounding system significant deficiencies include: disconnected grounding electrode conductor, loose, corroded, or other poor connection of the grounding electrode conductor to the grounding electrode, and grounding electrode conductor connection upstream from the service equipment.

7.1.A.3, 7.1.A.5 AND 7.1.A.7 – SERVICE EQUIPMENT AND ELECTRICAL PANEL INSPECTION REQUIREMENTS

The objective of the service equipment and electrical panel inspection is to identify and to report visual evidence that these components may not be functioning properly, or that they may be significantly deficient. These components can create an unsafe condition, and if one exists, the inspector should report the unsafe condition, usually as an implication of the improper functioning or significant deficiency that is causing the unsafe condition (see also Clause 2.2.1.B.1, unsafe). These components can have a limited service life, so end of service life reporting is required, when necessary.

In-scope service equipment and electrical panel components include the enclosures (cabinets), bus bar and terminal bars, overcurrent protection devices, conductors inside the enclosures, and conductors entering and leaving the enclosure.

The service equipment is often called the main disconnect, or a similar term. It may consist of up to six switches, fuses, or circuit breakers that when deactivated shut off all electric service to the building. The service equipment is usually, but not always, located in one enclosure. The service equipment enclosures may be inside or outside of the building. Electrical panels located downstream from the service equipment are subpanels.

Service Equipment and Electrical Panel Components Not Functioning Properly

The primary functions of service equipment and electrical panel components are to distribute electricity to branch circuits and to other electrical panels, and to protect the circuits and occupants from damage caused by anomalies such as excessive current, arcing, and ground-faults. If the circuits are supplied with electricity, then the service equipment and electrical panel components are usually functioning properly. It is possible for circuits to have electricity with improperly functioning service equipment and electrical panel components, but this situation is uncommon in residential buildings. Inspectors are not required to, and in most cases should not, operate or test service equipment and electrical panel components. It is, therefore, difficult to determine if these components are functioning properly during a home inspection, so determining this is out-of-scope (see also Clause 13.1.B.1 and 13.2.A.3). Refer to the discussion of Clause 7.1.A.9 for more about testing arc-fault circuit interrupters and ground-fault circuit interrupters.

Significantly Deficient Service Equipment and Electrical Panel Components

Some common examples of service equipment and electrical panel component significant deficiencies include: overcurrent device ampacity greater than conductor ampacity, too many conductors connected to a lug, deteriorated, damaged, and corroded components, improper or absent grounding and bonding connections, loose enclosure, overcurrent device manufacturer different from panelboard manufacturer, and known problem panels such as Federal Pacific.

Unsafe Service Equipment and Electrical Panel Components

Inspectors could consider any failure to operate properly or any significant deficiency as an unsafe condition if the deficiency creates an electrical shock hazard; for example, lack of an effective bonding connection creates an electrical shock hazard if a ground fault occurs. Service equipment and electrical panels with exposed energized parts are unsafe.

Service Equipment and Electrical Panel Components End of Service Life

Circuit breakers have an expected service life of around forty years. Inspectors should consider reporting circuit breakers if they appear to be more than forty years old.

Reporting end of service life for other service equipment and electrical panel components based on an arbitrary end of service life estimate is not required. If the components are in good condition, reporting end of service life is unnecessary. If the components are in poor condition, inspectors should report the poor condition, and report that the system may be near the end of its service life.

7.1.A.6 – ELECTRICAL CONDUCTOR INSPECTION REQUIREMENTS

The objective of the electrical conductor inspection is to identify and to report visual evidence that conductors may not be functioning properly, or that they may be significantly deficient. These components can create an unsafe condition, and if one exists, the inspector should report the unsafe condition, usually as an implication of the improper functioning or significant deficiency that is causing the unsafe condition (see also Clause 2.2.1.B.1, unsafe). Some conductor insulation and sheathing has a limited service life, so end of service life reporting is required, when necessary.

In-scope electrical conductors include all wiring methods such as cables, conduit, and tubing. Also in-scope are visible bonding connections, such as at metal water pipes, and bonding of metal raceways.

Bonding connections should be accessible for inspection; however, occasionally, they are not readily accessible, or they are not located where reasonably visible. Inspectors are not required to search for bonding connections that are not readily accessible, and those that are not reasonably visible. Inspectors should report if the metal water distribution pipe bonding connection is not located, and recommend evaluation to determine the location and condition of the bonding connection (see also Clauses 13.1.B.2.a, and 13.2.A.1).

Conductors Not Functioning Properly

The primary function of wiring methods is to safely and reliably conduct electricity to points of use, such as outlets. If the circuits are supplied with electricity, and there is no visible wiring method deterioration or damage, then the wiring method is usually functioning properly. It is possible for circuits to have electricity with an improperly functioning wiring method, but this situation is uncommon in residential buildings.

The primary function of bonding metal components that are not intended to be energized, but that could become energized, is to safely conduct the electricity to an overcurrent device that should open to stop current flow through the bonded metal components. Examples of metal electrical components that should be bonded include metal electrical appliance cabinets, all metal electrical service components, and metal electrical raceways.

It is not possible to determine if the bonding connections are functioning properly during a home inspection, so determining this is out-of-scope (see also Clause 13.1.B.1 and 13.2.A.3).

Significantly Deficient Conductors

Some common examples of wiring method and bonding connection significant deficiencies include: conductors installed where subject to physical damage, conductor splices not installed in a covered box, conductors improperly repaired with material such as electrical tape, deteriorated or damaged conductors including insulation and sheathing, inadequate conductor support, absent, disconnected, or loose bonding connections, and bonding connections on unclean surfaces such as those coated with rust or paint.

Some believe that all solid-conductor aluminum conductors are significantly deficient, but this belief is not universally accepted. Solid-conductor aluminum conductors installed from the middle 1960s to about the middle 1970s may be significantly deficient, and inspectors should consider reporting these conductors as such. Solid-aluminum conductors installed after the middle 1970s are probably not significantly deficient, but inspectors should report the presence of these conductors if required to do so by the standard of practice that they use. Reporting the presence of solid-conductor aluminum conductors is no longer required by the American Society of Home Inspectors (ASHI) Standard of Practice for Home Inspections (HISOP).

Unsafe Conductors

Inspectors may consider any failure to operate properly or any significant deficiency as an unsafe condition if the deficiency creates an electrical shock hazard. Conductors with exposed energized parts are unsafe.

Conductor End of Service Life

The service life of many wiring methods and all bonding connections should approximate the life of the building, if properly installed and maintained. Cloth-sheathed nonmetallic sheathed cable is considered to have a limited service life due to deterioration of the sheathing and of the insulation. Knob and tube wiring is considered to have a limited service life due to deterioration of the insulation. Inspectors in most markets should consider reporting the presence of these wiring methods and should consider reporting end of service life, when appropriate. Inspectors in some markets, such as Florida, should report cloth-sheathed nonmetallic sheathed cable and knob and tube wiring as being at end of service life because of homeowner insurance underwriting policies.

Reporting end of service life for other wiring methods and bonding connections based on an arbitrary end of service life estimate is not required. If the components are in good condition, reporting end of service life is unnecessary. If the components are in poor condition, inspectors should report the poor condition, and report that the system may be near the end of its service life.

7.1.A.8 – LIGHTS, SWITCHES, RECEPTACLES INSPECTION REQUIREMENTS

The objective of the lights, switches, and receptacles inspection is to identify and to report visual evidence that these components may not be functioning properly, or that they may be significantly deficient. These components can create an unsafe condition, and if one exists, the inspector should report the unsafe condition, usually as an implication of the improper functioning or significant deficiency that is causing the unsafe condition (see also Clause 2.2.1.B.1, unsafe). These components can have a limited service life; however, end of service life is based on component condition more than on age.

Inspection of all lights, switches, and receptacles is not required. Inspectors are required to inspect a representative number (a defined term) of these components. A representative number is one light, one switch, and one receptacle per room, and one of each component on each side of the building. Many inspectors inspect all such readily accessible components, but this is not required. (see also Clauses 13.1.B.2.a, and 13.2.A.1).

Inspection of lights includes ceiling fans. Inspection of lights and ceiling fans involves using the normal operating controls to determine if the light or ceiling fan activates and functions properly. The normal operating control for a light and ceiling fan is a switch or pull-chain. Only installed (a defined term) lights and ceiling fans are in-scope. Installed means attached to the building such that removal requires tools. Portable lights and fans are out-of-scope. Lights are occasionally controlled by timers or sensors that limit functioning to certain times or to certain conditions, such as darkness. Inspectors are not required to override or test these devices unless the controls are readily accessible and clearly labeled to indicate control function. Inspectors are not required to use or test remote controls to activate lights and ceiling fans unless the remote control is the only means to operate the light or ceiling fan, and if the remote control is readily accessible (see also Clause 7.2.A.1).

Some buildings are equipped with low-voltage lights. Inspection of low-voltage interior lights that use a point-of-use transformer to reduce the voltage at the light fixture is in-scope. Some low-voltage interior lighting systems, mostly from the 1960s, use transformers and relays at one or more central locations to control lights. Inspection of these systems is out-of-scope, as is inspection of exterior low-voltage lighting systems (see also Clause 7.2.A.3).

Inspection of switches includes three-way switches, four-way switches, dimmers, and fan speed controls. Each of these devices is a separate switch, and these switches are not covered by the representative number limitation.

Switches and other devices that serve as a required appliance or equipment disconnecting means are not switches addressed by the clause. Inspectors are required to visually inspect these devices, but they are not required to operate or test them.

Inspection of receptacles is visual; however, inspection of these components has evolved through common practice to include using a three-light receptacle tester. Inspectors should use a three-light tester to inspect 120-volt, three-slot receptacles. Inspectors may visually inspect 120-volt, two-slot receptacles and 240-volt receptacles. Inspectors are not required to use circuit analyzers and similar instruments to inspect receptacles (see also Clause 13.1.B.1).

Lights, Switches, Receptacles Not Functioning Properly

The primary function of a light is to activate and to provide illumination where intended. If the light activates, it is usually functioning properly. It is possible for a light to activate and not function properly, but this situation is very uncommon in residential buildings. Inspectors are not required to determine why a light does not activate, and should report failure to activate (see also Clause 13.2.A.4).

The primary function of a switch is to control the operation of electrical appliances and equipment, such as a light. If the appliance or equipment activates, the switch is usually functioning properly. It is possible for a light or appliance to activate with a switch that is not functioning properly, but this situation is uncommon in residential buildings. Three-way and four-way switches that operate appliances and equipment in one position but not in another position are not functioning properly.

Some switches control switched receptacles. Inspectors are not required to determine if a switch controls a receptacle. Determining what light or device a switch controls can be difficult. Inspectors are not required to determine what a switch controls if this is not readily apparent.

The primary function of a receptacle is to safely provide electricity to plug-and-cord connected appliances and equipment. If the appliance or equipment activates, the receptacle is often functioning properly. It is possible for a receptacle that is not functioning properly to provide electricity. This situation can be considered a significant deficiency.

Significantly Deficient Lights, Switches, and Receptacles

Some common examples of light and ceiling fan significant deficiencies include: damage and deterioration, loose fixtures, fixture too close to clothes closet storage area or to a bathtub or shower, light not installed where required, and ceiling fan wobbles or is unusually noisy.

Some common examples of switch significant deficiencies include: damage and deterioration, loose switches, absent or damaged cover plate, improper cover plate for switch location, switch not installed where required, improper operation of three-way and four-way switches, and inadequate support of switch box.

Some common examples of receptacle significant deficiencies include: damage and deterioration, painted receptacle, loose receptacle, absent or damaged cover plate, improper cover plate for receptacle location, improper wiring such as reverse polarity, ungrounded three-slot receptacle, receptacle not installed where required, and inadequate support of receptacle box.

Unsafe Lights, Switches, and Receptacles

Inspectors may consider any failure to operate properly or any significant deficiency as an unsafe condition if the deficiency creates an electrical shock hazard. Lights, switches, and receptacles with exposed energized parts are unsafe.

7.1.A.9 – GROUND-FAULT AND ARC-FAULT CIRCUIT INTERRUPTER INSPECTION REQUIREMENTS

The objective of the ground-fault circuit interrupter (GFCI) and arc-fault circuit interrupter (AFCI) inspection is to identify and to report visual evidence that these components may not be functioning properly, or that they may be significantly deficient. These components can create an unsafe condition, and if one exists, the inspector should report the unsafe condition, usually as an implication of the improper functioning or significant deficiency that is causing the unsafe condition (see also Clause 2.2.1.B.1, unsafe). These components have a limited service life; however, end of service life is based on component functionality more than on age.

Inspection of GFCI devices, including receptacles and circuit breakers, is visual; however, inspection has evolved over time to include testing these devices at the device using the manufacturer's test button. Inspectors are not required to determine whether receptacles downstream from the device are GFCI protected (see also Clause 13.2.A.8). Testing downstream receptacles for GFCI protection is generally not recommended because of the difficulty in finding and resetting GFCI receptacles that are concealed by occupant belongings or other materials. Inspectors are not required to move materials to locate GFCI receptacles, and are not required to inspect or to test GFCI receptacles that are not readily accessible (see also Clauses 13.1.B.2.a, 13.2.A.1, and 13.2.F.3).

Inspection of AFCI devices, including receptacles and circuit breakers, is visual. These devices are relatively new, and inspection techniques are evolving. Testing these devices in unoccupied buildings using the manufacturer's test button is the evolving inspection standard. Inspectors should test these devices in unoccupied buildings. Testing these devices in occupied buildings using the manufacturer's test button is at the inspector's discretion. Inspectors are not required to determine whether receptacles downstream from the device are AFCI protected (see also Clause 13.2.A.8). Testing downstream receptacles for AFCI protection is generally not recommended because of the difficulty in finding and resetting AFCI receptacles that are concealed by occupant belongings or other materials.

Inspectors are not required to move materials to locate AFCI receptacles, and are not required to inspect or to test AFCI receptacles that are not readily accessible (see also Clauses 13.1.B.2.a, 13.2.A.1, and 13.2.F.3).

Requirements for GFCI and AFCI protection of branch circuits, appliances, and equipment have evolved over time. Inspectors are not required to determine whether GFCI or AFCI protection is required on a branch circuit, appliance, or equipment (see also Clause 13.2.A.8).

GFCI and AFCI Devices Not Functioning Properly

The primary function of a GFCI device is to protect people from electric shock when metal that should not be energized is energized. The primary function of an AFCI device is to protect property from fire caused by arcing between wires. GFCI devices and AFCI devices are completely different, and may not be substituted for each other. Circuit breakers that combine these devices are available.

It is not possible to determine if a GFCI device or an AFCI device is functioning properly without using the manufacturer's test button. Newer GFCI devices should discontinue functioning if the device is not functioning properly. Refer to the previous discussion regarding testing these devices.

Significantly Deficient GFCI and AFCI Devices

Some common examples of GFCI and AFCI significant deficiencies include: damage and deterioration.

Unsafe GFCI and AFCI Devices

Inspectors may consider GFCI and AFCI devices that are not functioning properly as being unsafe. Inspectors may consider recommending installing GFCI and AFCI devices as an upgrade in buildings where these devices were not required when the building was built.

7.1.B.1, 7.1.B.2, 7.1.B.3, 7.1.B.4 – ELECTRICAL SYSTEM DESCRIPTIONS

These clauses require the inspector to describe: the service amperage rating, location of the service equipment (main disconnect) and all panels including subpanels, whether smoke alarms and carbon monoxide alarms are present, and the predominant branch circuit wiring method. As previously discussed in Clause 2.2.B.1, a description should allow the reader to distinguish the described materials, equipment, and locations from similar materials, equipment, and locations.

Service Amperage Description

The building's service amperage rating is determined by the component with the lowest amperage rating beginning at the service entrance conductors, then the meter base, the panelboard, and the service equipment. In many, but not all, cases, the service equipment amperage rating determines the service amperage rating. Situations exist where it is not possible for the home inspector to determine the service amperage rating. These situations are most likely to occur in older buildings where the service amperage has been increased, or where the service components have been altered, often without a permit. Inspectors should report if the service amperage rating was not determined and recommend electrician evaluation to determine the service amperage rating.

Typical Panel Location Descriptions

Describing the precise location of the service equipment, panels, and subpanels is not required; a general location is acceptable. Directions such as right, left, front, and back are acceptable if the reference point is defined in the report, such as at the street looking at the front of the house.

The following descriptions are examples of panel location descriptions, and are not intended to be a complete list of all such descriptions. The service equipment (main disconnect) is located at the right side exterior of the building. The building is served by a subpanel located near the air conditioning condensers in the left side yard.

Smoke Alarm and Carbon Monoxide Alarm Descriptions

Inspectors are only required to report whether smoke alarms and carbon monoxide alarms are present, or to report if either alarm is not present. Inspectors are not required to inspect or to test these alarms (see also Clause 7.2.A.2). Inspectors are not required to determine the age or type of these alarms, such as ionization or photovoltaic (see also Clause 7.2.C). Inspectors are not required to determine if the alarms are installed where they are currently required, and are not required to determine if the alarms are properly installed, including interconnection (see also Clause 13.2.A.8). Some inspectors perform these services, but this is not required. Inspectors may consider recommending installing or upgrading these alarms to improve safety, but this is not required.

Some inspectors inspect and test these alarms, but this is not required. Inspectors should take care to limit the expectations of these procedures. Inspectors should not state that inspecting or testing these alarms indicates that they are functioning properly.

Predominant Wiring Method Descriptions

Wiring methods include cables such as nonmetallic sheathed cable (often referred to by the brand name Romex), and various types of metallic and nonmetallic conduit and tubing. Nonmetallic sheathed cable is the predominant wiring method in most single-family residential buildings in most markets. In older buildings, approximately 1930s and older, and in some newer buildings, inspectors often encounter multiple wiring methods. Inspectors are only required to describe the wiring method that the inspector believes accounts for fifty percent or more of the wiring in the building. The inspector may describe additional wiring methods, but this is not required.

The following descriptions are examples of wiring method descriptions, and are not intended to be a complete list of all such methods. Typical wiring method descriptions include: armored cable (AC) (sometimes referred to by the brand name BX), electrical nonmetallic tubing (ENT), intermediate metallic conduit (IMC), and nonmetallic sheathed cable (NM).

7.2.A.1 – REMOTE CONTROL INSPECTION LIMITATION

Inspectors are not required to locate, use, or test the remote control for most types of appliances and equipment. This includes electrical appliances such as ceiling fans, and HVAC appliances such as some mini-split systems.

Some appliances and equipment can only be operated using a remote control. Inspectors should use the remote control as the normal operating control for such appliances and equipment. Inspectors should report if the remote control is not visible or is not readily accessible, and should recommend that the seller produce the remote control and demonstrate its operation.

7.2.A.3 AND 7.2.A. 4 – LOW VOLTAGE AND ANCILLARY WIRING INSPECTION LIMITATION

Low voltage is usually defined as being less than thirty volts. Common examples of low voltage wiring include doorbells, some older thermostats, and some interior and exterior lighting systems.

Common examples of ancillary wiring systems include telephone, television, computer, and intercom. Ancillary wiring systems also include backup generators and associated wiring and transfer switches.

Inspectors are not required to inspect these wiring systems, or to inspect or operate devices that are part of or are served by these systems. Many inspectors test doorbells, but this is not required. Inspectors should consider reporting the presence of systems such as backup generators and low voltage lighting, disclaiming inspection, and recommending inspection by a qualified specialist.

7.2.A.5 – RENEWABLE ENERGY SYSTEM INSPECTION LIMITATION

Inspectors are not required to inspect, or to operate, systems such as solar photovoltaic and wind energy systems, including all associated equipment and wiring. Inspectors should consider reporting the presence of these systems, disclaiming inspection, and recommending inspection by a qualified specialist.

7.2.B – ELECTRICAL MEASUREMENT LIMITATION

Inspectors are not required to measure any electrical system parameters including current, voltage, and impedance (resistance) (see also Clause 13.1.B.1).

7.2.C - SMOKE ALARM AND CARBON MONOXIDE ALARM AGE AND TYPE LIMITATION

Refer to Clause 7.1.B.4 for the discussion of this limitation (see also Clause 7.2.A.2).

HISOP Sections 8 and 9 - Heating and Cooling

INTRODUCTION

These sections contain the requirements and specific limitations for inspecting heating and air conditioning (cooling) system components including: installed heating and cooling equipment (appliances), vent systems and chimneys that serve the heating system, normal operating controls (thermostats), refrigerant line sets, condensate disposal systems, and heating and cooling distribution systems, including typical return air filters.

8.1.A AND 9.1.A – ACCESS PANEL OPENING REQUIREMENTS

Refer to Clause 2.2.A for the discussion of these requirements.

8.1.B.1 AND 9.1.B.1 – HEATING AND COOLING APPLIANCE INSPECTION REQUIREMENTS

The objective of the heating and cooling equipment (appliance) inspection is to identify and to report visual evidence that the appliance may not be functioning properly, or that it may be significantly deficient. These components can create an unsafe condition, and if one exists, the inspector should report the unsafe condition, usually as an implication of the improper functioning or significant deficiency that is causing the unsafe condition (see also Clause 2.2.1.B.1, unsafe). These appliances have a limited service life, so end of service life reporting is required, when appropriate.

In-scope heating appliances include installed appliances that are intended to provide the required heat to the building. Most residential buildings are required to have one or more installed heating appliances that together can maintain the indoor temperature at a minimum of 68° F. Examples of in-scope heating appliances include: fuel-fired central furnaces and boilers, such as those using gas, oil, or wood, electric central furnaces and boilers, fuel-fired floor furnaces, wall furnaces, installed room heaters, electric baseboard heaters, and electric radiant heating strips and panels. Examples of out-of-scope appliances and systems that provide heat include solid-fuel-burning fireplaces and stoves, and gas-fired fireplaces (technically decorative gas appliances). These appliances are addressed in Section 12 of the American Society of Home Inspectors (ASHI) Standard of Practice for Home Inspections (HISOP). Heating appliances that are not installed, such as portable heaters, are out-of-scope, even if they are supplied with fuel from the building fuel distribution system.

In-scope cooling appliances include installed appliances that are intended to provide cooling in the building. Cooling is not required; however, inspectors should report the absence of installed cooling appliances in markets where cooling is common and expected. Examples of in-scope cooling appliances include: evaporator coils and condensers installed with forced air central furnaces, appliances installed through an opening in a wall, evaporative coolers, and mini-split systems. Examples of out-of-scope appliances include appliances installed in a window, and portable evaporative coolers.

The normal operating control for most heating and cooling systems is a thermostat. Inspectors should use the normal operating controls to operate the systems, and should use no other method to operate the systems. Inspectors are not required to operate complex programmable thermostats and internet-connected thermostats if operating instructions are not readily accessible (see also Clauses 13.2.D.1 and 13.2.F.5). Inspection of thermostats is limited to observing the condition of the thermostat. Inspectors are not required to determine the accuracy of thermostats (see also Clause 13.2.A.3).

Cooling systems and heat pumps with an evaporator coil indoors and a condenser outdoors have a refrigerant line set to circulate refrigerant between the evaporator coil and the condenser. Inspection of the refrigerant line set is limited to observing the condition of the line set such as reasonable line set support, and the presence and condition of insulation on the suction (larger) tube.

Cooling systems, and high efficiency heating systems, produce water as a byproduct of operating. The condensate disposal system captures this water and conducts it to a place for disposal, usually, but not always, outside the building. Inspection of this system is limited to a visual inspection of component presence and condition. Inspectors are not required to test or operate components such as condensate pumps and condensate overflow cutoff switches (see also Clause 13.2.C.4).

Inspectors are not required to operate heating and cooling systems if the inspector believes that doing so may damage the system (see also Clause 13.2.D.1). Operating cooling systems at a low outside air temperature is likely to provide an inaccurate result regarding whether the system is functioning properly, so operating cooling systems at low outside air temperature may not be worthwhile regardless of the risk.

Heating systems, such as boilers, are sometimes shut down outside of heating season. Some heating system controls have limits on system operation during warm weather. Inspectors are not required to determine the reason why a heating or cooling system does not appear to be functioning properly, nor are inspectors required to activate a system that is shut down (see also Clauses 13.2.A.4, 13.2.C.1, and 13.2.C.2). Inspectors should report if the heating or cooling system was not operated, report the reason why, and recommend evaluation of the system (see also Clause 2.2.B.4).

Heating and Cooling Appliances Not Functioning Properly

The primary function of heating appliances and their distribution systems is to safely maintain the temperature inside the conditioned areas of the building at not less than 68° F. The primary function of cooling appliances and their distribution systems is to maintain the temperature inside the conditioned areas of the building at the assumed design temperature. The ability of a heating or cooling system to maintain the interior temperature is based on several design parameters that include an assumed outside air temperature. It is often not possible to operate a heating or cooling system at the assumed outside air temperature, so it is often not possible to determine if the heating or cooling appliance can perform its primary function; thus, determining whether a heating or cooling system will maintain proper indoor temperature is out-of-scope (see also Clause 13.2.A.3). It is also out-of-scope to determine heating or cooling system temperature balance and distribution adequacy in individual rooms and areas of the building (see also Clauses 8.2.B.1 and 13.2.A.3).

Some common examples of heating and cooling appliances not functioning properly include: failure to activate using normal operating controls, or failure to fully activate using normal operating controls, improper flame color or shape, inadequate temperature difference between air entering and leaving the evaporator coil (may or may not be a defect), and ice on cooling system components.

Many inspectors measure the temperature difference between air entering an evaporator coil and air leaving an evaporator coil. Some inspectors measure condenser current draw. Performing these measurements is out-of-scope (see also Clause 13.1.B.1). These measurements may not provide an accurate indication of whether the system is functioning properly. Inspectors should exercise care when reporting the results of these measurements to avoid misunderstandings about the implications of these measurements.

Heating and cooling appliances, such as floor furnaces, wall furnaces, and evaporative coolers are sometimes abandoned when another heating or cooling system is installed. Abandoned systems should be removed; however, they frequently are not. Inspectors should report the presence of abandoned systems and recommend removal.

Significantly Deficient Heating and Cooling Appliances

Some common examples of heating and cooling appliance significant deficiencies include: damaged, deteriorated, or inoperative parts, debris or rust in cabinet, debris on evaporator coil or on condenser fins, appliance not level, inadequate clearance to combustible materials, visible refrigerant stains, condensate flow from secondary condensate disposal pipe, water or water stains in auxiliary condensate pan, and inadequate service clearances.

Unsafe Heating and Cooling Appliances

Inspectors may consider heating and cooling appliances that fail to function properly or have significant deficiencies to be unsafe if the deficiency creates a significant risk of injury. Examples may include deficiencies that could produce carbon monoxide and deficiencies that could create a fire hazard.

Heating and Cooling Appliances End of Service Life

The cost to replace heating and cooling appliances is significant, so clients are often concerned about the appliances, and about their remaining service life. Inspectors are not required to, and should not, speculate about the remaining service life of any system or component (see also Clause 13.2.A.2). Inspectors are required to report about systems and components that appear near the end of their service life. Refer to the discussion of Clause 2.2.B.1 for more discussion about end of service life reporting.

The service life of a heating and cooling appliance is a function of the appliance type, quality, environment, maintenance, and installation. The actual service life of these appliances varies widely. Inspectors should report heating and cooling appliances that appear to be near the end of their service life based on appliance age (if determined) and on physical condition. Note that determining appliance age is not required, and is sometimes not possible.

8.1.B.2 – VENT SYSTEM INSPECTION REQUIREMENTS

Refer to Clause 6.1.A.4 for a discussion of these requirements.

8.1.B.3 AND 9.1.B.2 – DISTRIBUTION SYSTEM INSPECTION REQUIREMENTS

The objective of the heating and cooling distribution system inspection is to identify and to report visual evidence that the system may not be functioning properly, or that it may be significantly deficient. These systems rarely create an unsafe condition, but if one exists, the inspector should report the unsafe condition, usually as an implication of the improper functioning or significant deficiency that is causing the unsafe condition (see also Clause 2.2.1.B.1, unsafe). The service life of these systems, if properly installed and maintained, should approximate the life of the building. End of service life is based on component condition more than on age.

Whether heating and cooling distribution system components are in-scope depends on the heating system, and on whether the components are visible and readily accessible (see also Clauses 13.1.B.2.a and, 13.2.A.1). Heating and cooling distribution system components are often concealed in walls, ceilings, and floors, and are often located in inaccessible areas of attics and crawlspaces.

In-scope heating and cooling distribution systems include low-pressure forced-air ducts, and pipes and tubes for hydronic heating systems (hot water and steam). Out-of-scope heating and cooling distribution systems include high-pressure forced-air ducts because installation of these systems is based on manufacturer's instructions (see also Clause 13.2.A.8). These instructions vary widely between manufacturers and between models from the same manufacturer. Inspectors are not required to determine compliance with manufacturer's instructions (see also Clause 13.2.A.8). Inspectors should inspect for visual evidence that these high-pressure systems are not functioning properly, but confirming proper installation of these systems is out-of-scope.

Distribution System Not Functioning Properly

Some common examples of heating and cooling distribution system not functioning properly include: poor air flow from ducts, low heat or no heat at hydronic system distribution devices, and malfunctioning air valves and vents in steam systems.

Significantly Deficient Distribution System

Some common examples of heating and cooling distribution system significant deficiencies include: air leaks from ducts, water or steam leaks from pipes, vents, and traps, crimped ducts, improperly supported ducts and pipes, sagging ducts, deteriorated or damaged components, significant and abnormal condensation on ducts, flexible ducts located in a garage, supply or return openings in a garage, and dirty or clogged filters.

8.1.C.1, 8.1.C.2, AND 9.1.C.1, 9.1.C.2 - HEATING AND COOLING SYSTEM DESCRIPTIONS

These clauses require the inspector to describe the types of heating and cooling systems serving the building and their energy sources. As previously discussed in Clause 2.2.B.1, a description should allow the reader to distinguish the described systems and energy sources from similar systems and energy sources. If multiple types of systems and energy sources are present, the inspector should describe each type of system and energy source.

Typical Heating and Cooling Appliance and Energy Source Descriptions

Describing the heating and cooling appliance capacity is not required. Describing the location of heating and cooling appliances is not required. Describing the model number and serial number is not required. Describing the distribution system is not required. Describing the type of fuel-gas is not required; however, inspectors may consider describing this if the fuel is propane, especially if there is no visible propane tank.

The following descriptions are examples of heating and cooling system and energy source descriptions, and are not intended to be a complete list of all such descriptions. Heat and air conditioning for the first story is provided by a gas-fired furnace and evaporator coil and an air-source condenser. Heat is provided by an oil-fired hot water boiler. Heat for the sunroom is provided by electric baseboard heaters. Heat and air conditioning for the rear room is provided by a mini-split system.

8.2.A.1 – FLUE INTERIOR INSPECTION LIMITATION

Refer to Clause 5.2.B for the discussion of this limitation.

8.2.A.2 – HEAT EXCHANGER INSPECTION LIMITATION

Heat exchangers in modern heating systems are rarely readily accessible and visible for inspection without dismantling. Even when a heat exchanger is visible, visibility is usually very limited. Inspecting heat exchangers is, therefore, out-of-scope. Use of alternative inspection methods (such as a carbon monoxide detector) that may detect a heat exchanger defect, are also out-of-scope (see also Clause 13.1.B.1 and 13.2.A.1).

8.2.A.3 – HUMIDIFIER AND DEHUMIDIFIER INSPECTION LIMITATION

Inspecting humidifiers and dehumidifiers is out-of-scope. This includes installed appliances that are connected to installed heating and cooling systems, and portable appliances that are not installed and that are not connected to installed heating and cooling systems.

Installed humidifiers are often poorly maintained and are often not functioning properly. In addition, the presence of dehumidifiers can indicate a moisture-related problem in the building. Inspectors should consider reporting the presence of these appliances, if visible, disclaiming inspection, and recommending specialist evaluation if the client wants assurance about the installation, condition, and functionality of these systems.

8.2.A.4 AND 9.2.A – AIR CLEANING AND SANITIZING DEVICE INSPECTION LIMITATION

Inspecting air cleaning and sanitizing devices is out-of-scope. This includes installed devices that are connected to installed heating and cooling systems, and portable devices that are not installed and that are not connected to installed heating and cooling systems. Examples of these devices include electrostatic air filters and ultraviolet lights.

It is not possible for inspectors to determine if these devices are functioning properly. Inspectors should consider reporting the presence of these devices, if visible, disclaiming inspection, and recommending specialist evaluation if the client wants assurance about the installation, condition, and functionality of these devices.

8.2.A.5 AND 9.2.D – ALTERNATIVE ENERGY EQUIPMENT INSPECTION LIMITATION

Refer to Clause 6.2.A.5 for the discussion of this limitation.

8.2.A.6 – HEAT RECOVERY SYSTEM AND ENERGY RECOVERY SYSTEM INSPECTION LIMITATION

Inspecting heat recovery ventilation systems, energy recovery ventilation systems, and other central exhaust and ventilation systems is out-of-scope. These systems can improve the efficiency of forced-air heating and air conditioning systems by using exhaust air from inside the building to heat or cool incoming ventilation air, and by removing moisture from the air during cooling season. These systems can improve indoor air quality. Efficiency and effectiveness of these systems depends on how they are installed and maintained.

These systems must be maintained in order to operate properly. Inspectors may consider opening a readily openable access panel to determine if the system has been properly maintained, but this is not required.

Proper operation of these systems depends on the quality of the installation, and on proper maintenance. It is not possible for inspectors to determine if these systems are functioning properly. Inspectors should consider reporting the presence of these systems, if visible, disclaiming inspection, and recommending specialist evaluation if the client wants assurance about the installation, condition, and functionality of these systems.

8.2.B.1 AND 9.2.B – HEAT AND COOLING ADEQUACY AND BALANCE INSPECTION LIMITATION

Heating and cooling loads, system capacity, and distribution system design are specified in publications such as the Air Conditioning Contractors of America Manual J, Manual S, and Manual D. Contractors are responsible for selecting and using the appropriate design tools before installing heating and cooling appliances and distribution systems. It is out-of-scope for inspectors to determine if appropriate design specifications were used (see also Clauses 13.1.B.1 and 13.2.A.3).

So-called rules-of-thumb for estimating heating and cooling appliance capacity can be significantly inaccurate. Inspectors are not required to, and in most cases should not, attempt to apply these rules-of-thumb.

8.2.B.2 – COMBUSTION AIR ADEQUACY INSPECTION LIMITATION

Determining whether combustion air ducts or openings are adequate to provide combustion air to fuel-fired appliances requires calculations, and is, therefore, out-of-scope (see also Clauses 13.1.B.1, 13.2.A.3, and 13.2.A.8).

Inspection of combustion air ducts or openings consists of determining if such ducts or openings are present and in acceptable condition. Inspectors should also inspect for visible indications of inadequate combustion air, such as soot and rust stains at combustion appliances and at vent terminations.

HISOP Section 10 - Interiors

INTRODUCTION

This section contains the requirements and specific limitations for inspecting components inside the building including: walls, ceilings, floors, steps, stairs, handrails, guards, counter-tops, cabinets, doors, windows, vehicle doors and their operators, and installed appliances.

It is important to distinguish between functional deficiencies and cosmetic deficiencies, especially when inspecting components inside the building. Cosmetic deficiencies are those that do not significantly affect a component's performance of its intended function. Inspectors are not required to inspect for, or to report about, cosmetic deficiencies (see also clause 13.1.B.2.b). Home inspections provide more benefits to clients when the inspector focuses on functional deficiencies, and focuses on areas where people usually do not go, such as attics and crawlspaces.

10.1.A – WALL, CEILING, AND FLOOR INSPECTION REQUIREMENTS

The objective of the wall, ceiling, and floor inspection is to identify and to report visual evidence that these components may not be functioning properly, or that they may be significantly deficient. These components rarely create an unsafe condition, but if one exists, the inspector should report the unsafe condition, usually as an implication of the improper functioning or significant deficiency that is causing the unsafe condition (see also Clause 2.2.1.B.1, unsafe). The service life of these components, if properly installed and maintained, should approximate the life of the building. End of service life is based on component condition more than on age.

Inspection of interior walls, ceilings, and floors sometimes involves observing and interpreting indications of significant deficiencies involving the wall, ceiling, and floor coverings themselves, such as drywall and plaster. More often, however, inspection involves observing and interpreting indications of improper functioning and significant deficiencies attributable to other systems and components. For example, some cracks in interior wall and ceiling coverings, such as drywall and plaster, can be the result of improper installation and finishing of the wall and ceiling coverings. These are examples of significant deficiencies of the coverings themselves. Other cracks can be the result of structural deficiencies. Floors that are not level may be the result of structural deficiencies, or they may have been built that way. Water stains are usually the result of significant deficiencies of plumbing components, flashing, windows or doors, or roof coverings. Inspectors are not required to, and in most cases should not attempt to, determine the cause of deficiencies (see also Clause 13.2.A.4). Inspectors should report the deficiency indication and recommend evaluation to determine cause and recommended actions (see also Clause 2.2.B.2).

It is important to distinguish between interior wall, ceiling, and floor coverings and the decorative coatings, decorative coverings, and other finish treatments applied to the wall, ceiling, and floor. Inspectors are not required to inspect, or to report about, deficiencies of: paint, stain, and wallpaper, floor coverings such as carpet, wood, laminates, and vinyl, and window treatments such as drapes, blinds, and shutters (see also Clauses 10.2.A, 10.2.B, 10.2.C, and 13.2.E.2).

Some ceilings, usually in basements, consist of panels installed in frames that are suspended from framing. These are usually called drop ceilings or suspended ceilings. These panels can be difficult to replace in the grid once moved, and can be damaged when moved. Matching replacements can be difficult or impossible to acquire. Some inspectors move these panels, but moving these panels is not required (see also Clause 13.2.F.1).

Walls, Ceilings, and Floors Not Functioning Properly

The primary function of interior wall, ceiling, and floor coverings is to separate conditioned areas from unconditioned areas. These coverings may also have some structural value, and may be part of a vapor retarder system. Inspectors are not required to determine if walls, ceilings, and floors are serving as structural elements, or as vapor retarders (see also Clauses 13.1.B.1, 13.2.A.3, and 13.2.B.2).

Some common examples of walls, ceilings, and floors not functioning properly include: protruding fasteners (nail pops), visible openings (holes), and some types of cracks.

Significantly Deficient Walls, Ceilings, and Floors

Some common examples of wall, ceiling, and floor significant deficiencies include: damage or deterioration, and some types of cracks.

10.1.B – STEPS, STAIRS, AND RAILINGS INSPECTION REQUIREMENTS

The objective of the steps, stairs, and railings inspection is to identify and to report visual evidence that these components may not be functioning properly, or that they may be significantly deficient. These components often create an unsafe condition, and if one exists, the inspector should report the unsafe condition, usually as an implication of the improper functioning or significant deficiency that is causing the unsafe condition (see also Clause 2.2.1.B.1, unsafe). The service life of these components, if properly installed and maintained, should approximate the life of the building. End of service life is based on component condition more than on age.

The term railings includes handrails and guards. Guards are fall-protection components that are required when a walking surface is more than thirty inches above another surface. A guard may be a railing, but it may also be a wall or any other structure that complies with the guard height and load-bearing requirements.

Steps, Stairs, and Railings Not Functioning Properly

The primary functions of these components include: supporting vertical and horizontal design loads, providing a safe path while using these components, and preventing falls. Some common examples of steps, stairs, and railings not functioning properly include these components being loose or subject to excessive deflection.

Significantly Deficient Steps, Stairs, and Railings

Some common examples of steps, stairs, and railings significant deficiencies include: uneven riser height and tread depth, excessive riser height, inadequate tread depth, inadequate ceiling height, absent guard, inadequate guard height, absent handrail, and handrail not graspable.

Unsafe Steps, Stairs, and Railings

Step, stair, and railing standards have changed over time. Lack of compliance with current standards does not necessarily make these components unsafe. Inspectors may consider reporting lack of compliance with current standards as an unsafe condition, or inspectors may consider recommending upgrading these components to improve safety (see also Clause 2.2.B.1).

10.1.C – COUNTERTOPS AND CABINETS INSPECTION REQUIREMENTS

The objective of the countertop and cabinet inspection is to identify and to report visual evidence that these components may not be functioning properly, or that they may be significantly deficient. These components rarely create an unsafe condition, but if one exists, the inspector should report the unsafe condition, usually as an implication of the improper functioning or significant deficiency that is causing the unsafe condition (see also Clause 2.2.1.B.1, unsafe). These components have a limited service life; however, end of service life is based on component condition more than on age.

Inspection of all countertops is required. Inspection of all cabinets is not required. Inspectors are required to inspect a representative number of installed cabinets, including their doors and drawers. A representative number is one drawer and one door per cabinet. Many inspectors inspect all cabinets and their doors and drawers, but this is not required. (see also Clauses 13.1.B.2.a, and 13.2.A.1).

Countertops and Cabinets Not Functioning Properly

The primary function of countertops is to provide a stable, level, and sanitary surface. The primary function of cabinets is to provide a secure place to store belongings. Some common examples of cabinets and countertops not functioning properly include being significantly out of level and being insecurely attached to the wall or floor. Note that some island cabinets in kitchens are intended to be moved.

Significantly Deficient Countertops and Cabinets

Some common examples of cabinet and countertop significant deficiencies include: significant deterioration or damage, loose doors and drawers, poor door and drawer operation, and significantly sagging shelves.

10.1.D – WINDOWS AND INTERIOR DOORS INSPECTION REQUIREMENTS

The objective of the window and interior door inspection is to identify and to report visual evidence that these components may not be functioning properly, or that they may be significantly deficient. These components rarely create an unsafe condition, but if one exists, the inspector should report the unsafe condition, usually as an implication of the improper functioning or significant deficiency that is causing the unsafe condition (see also Clause 2.2.1.B.1, unsafe). These components have a limited service life; however, end of service life is based on component condition more than on age.

Inspection of all exterior doors is required (see also Clause 4.1.A.2). Inspection of all windows and interior doors is not required. Inspectors are required to inspect a representative number of windows and interior doors. A representative number is one window and one interior door per room. Most inspectors inspect all windows and all interior doors, but this is not required (see also Clauses 13.1.B.2.a, and 13.2.A.1).

Windows and Interior Doors Not Functioning Properly

The primary functions of windows are to admit light, admit air when opened, keep out air and water when closed, and provide a minimal level of security. The primary function of interior doors is to open, close, and latch without rubbing, sticking, or binding. Some common examples of windows not functioning properly include: not opening easily, not remaining open, not closing completely, damaged glazing, and water stains that indicate water infiltration around or through the window. Some common examples of interior doors not functioning properly include: rubbing, sticking or binding, not remaining open, and not latching.

Significantly Deficient Windows and Interior Doors

Some common examples of window and interior door significant deficiencies include: deterioration, damage, and sash locks, cranks, and other hardware absent or damaged.

Unsafe Windows and Interior Doors

Refer to Clause 2.2.B.1, unsafe, for a discussion of this topic.

10.1.E – VEHICLE DOOR AND VEHICLE DOOR OPERATOR INSPECTION REQUIREMENTS

The objective of the vehicle door and the vehicle door operator inspection is to identify and to report visual evidence that these components may not be functioning properly, or that they may be significantly deficient. These components can create an unsafe condition, and if one exists, the inspector should report the unsafe condition, usually as an implication of the improper functioning or significant deficiency that is causing the unsafe condition (see also Clause 2.2.1.B.1, unsafe). These components have a limited service life; however, end of service life is based on component condition more than on age.

Vehicle doors are not always located in a garage. For example, a vehicle door could be located in a basement. This clause applies to all vehicle doors regardless of location in the building being inspected. Vehicle doors located in buildings not being inspected are out-of-scope. Note, however, that detached garages are in-scope (see also Clause 13.1.C).

Vehicle doors are heavy, some can weigh up to 400 pounds. These doors require the use of springs under tension to assist with door operation. These springs have a limited service life, and can wear out or become improperly adjusted over time. This condition is often not noticeable when the door is connected to a vehicle door operator. Inspectors should consider disconnecting the door from the vehicle door operator to inspect the operation of the door and condition of the springs, but this is not required.

Testing the vehicle door operator pressure safety reverse function is not required (see also Clause 13.2.C.4); however, some inspectors test this function, and testing this function is required by some other standards of practice. The only approved method of testing this function is to close the vehicle door on a 1 ½ inch thick solid object. Other tests, such as using hands to catch the vehicle door near mid-closing, are not valid, and inspectors should not report the results of these invalid tests. The approved test, however, subjects the vehicle door to a risk of damage. Inspectors are not required to perform procedures that are subject to this risk (see also Clause 13.2.F.1). Inspectors should consider reporting if the pressure safety reverse test was not performed, and recommending evaluation of the vehicle door and operator.

Vehicle Door and Operator Not Functioning Properly

The primary functions of these components include: separating the garage from the outside, opening and closing manually with reasonable effort, and providing some security. Some common examples of garage vehicle door and operator not functioning properly include: unusual noise while operating, binding during operation, door not remaining open when opened to near the center of the opening, operator not reversing when striking a 1 ½ inch thick object.

Significantly Deficient Vehicle Door and Operator

Some common examples of vehicle door and operator significant deficiencies include: damage, deterioration, absent and loose parts, and operator sensors or wall switch improperly installed.

Unsafe Vehicle Door and Operator

Some old operators without sensors are still in use. Inspectors may consider reporting these operators both as unsafe and as past the end of their service lives. Springs without containment cables are considered unsafe.

10.1.F – APPLIANCE INSPECTION REQUIREMENTS

The objective and scope of the appliance inspection is limited. The objective of the appliance inspection is to determine whether the primary appliance function is functioning properly. Inspectors should consider reporting significant deficiencies, visible unsafe conditions, and appliances near end of service life, but this is not required. (see also Clauses 10.2.H, 10.2.I, and 13.1.A).

The primary appliance function is the function that the occupant is likely to use most often. For an oven, this is the bake function using the lower heating component. For a range, this is the operation of all surface burners and the oven bake function using the lower heating component. For a surface cooking appliance, this is the operation of all burners. For a microwave oven, it is the timed cooking function. For a dishwashing machine, it is any function that causes the machine to input water, operate, and drain water. For a food waste grinder, it is operation using the normal operating control.

The scope of the appliance inspection is further limited by other clauses. Operation and testing of other appliance functions and features, such as timers, clocks, and self-cleaning cycles is out-of-scope (see also Clauses 10.2.H and 10.2.I).

Only installed appliances listed in Clause 10.1.F are in-scope. Appliances, such as refrigerators, freezers, wine coolers, trash compactors, and laundry appliances are out-of-scope. (see also Clauses 10.2.G, and 13.1.A). Exhaust systems for cooking areas, such as exhaust hoods and downdraft exhaust systems, are addressed in Section 11 of the American Society of Home Inspectors (ASHI) Standard of Practice for Home Inspections (HISOP).

Appliances are sometimes not located in the kitchen. In-scope appliances that are located inside a building being inspected are in-scope and should be operated. Appliances that are not located in a building being inspected, or that are located outside, are out-of-scope (see also Clauses 13.2.E.5 and 13.2.E.8).

Appliances Not Functioning Properly

Some common examples of appliances not functioning properly include: cooking element does not activate, or does not fully activate, microwave oven does not operate, or does not heat the test material, dishwashing machine does not fill, operate, fully drain, or leaks, food waste grinder does not operate, or leaks.

10.2.A, B, AND C – INTERIOR WALL COATINGS, WALL COVERINGS, FLOOR COVERINGS, AND WINDOW TREATMENTS INSPECTION LIMITATION

Inspectors are not required to report about the condition of wall coatings, such as paint and stain, and wall coverings, such as wallpaper. This includes cosmetic deficiencies on walls, ceilings, and trim molding. This includes the condition of caulk when the caulk serves only a cosmetic purpose (see also Clause 13.1.B.2.b).

Inspectors are not required to report about the condition of any floor covering, such as carpet, wood, tile, laminate, and vinyl. Significant deficiencies in floor coverings can indicate significant deficiencies in other systems and components. For example, cracks through the tiles of tile floor coverings can indicate a reportable structural deficiency.

Inspectors are not required to report about the condition of any window treatments such as drapes, blinds, and shutters. Inspectors are not required to operate window treatments if the inspector believes that doing so may damage the window treatment, or may cause it to become dislodged (see also Clause 13.2.F.1). Inspectors should report if the window treatment is not operated and if this restricts inspection of the window.

Inspectors may consider reporting about the condition of these out-of-scope components, especially if the condition is unusually poor, however, this is not required. Inspectors should report about the condition of walls, ceilings, and floors if a significant deficiency exists that is likely caused by another system or component not functioning properly, or being significantly deficient (see also Clause 10.1.A).

10.2.D – WINDOW COATINGS AND SEALS INSPECTION LIMITATION

Most windows in modern buildings are constructed with two or more panes of glazing with space between the panes. This space is filled with air in most windows, or with an inert gas in more energy efficient windows. The perimeter of the glazing is sealed to prevent air infiltration between the panes. This seal usually deteriorates over time allowing air infiltration between the panes. This condition is a significant defect for which there may be no practical repair; sash or glazing replacement is the usual option. Evidence of this condition usually presents as water stains or fogging between the panes. Sometimes this condition is clearly visible, but often it is not. Inspectors are not required to report about the condition of window glazing seals. Inspectors should consider reporting visible indications of glazing seal failure because the cost of replacing failed window sashes can be significant, but this is not required.

Some windows have a low emissivity coating that reflects infrared radiation. This coating can sometimes be seen, but sometimes a device is required to detect the presence of this coating. The proper location of this coating on the window depends on the climate where the building is located. Inspectors are not required to report about the presence, installation, or condition of this or any other window coating (see also Clause 13.1.A.3).

10.2.E – CENTRAL VACUUM SYSTEM INSPECTION LIMITATION

Inspectors are not required to inspect central vacuum systems including motors, tubes, and accessories. Inspectors should consider reporting the presence of these systems, disclaiming inspection, and recommending specialist evaluation if the client wants information about the installation and condition of these systems.

10.2.F – RECREATIONAL FACILITIES INSPECTION LIMITATION

Inspectors are not required to inspect recreational facilities (a defined term). Examples of recreational facilities include swimming pools, spas, saunas, steam baths, playground equipment, play structures, and athletic areas such as tennis courts and basketball goals. Inspectors should consider reporting the presence of these recreational facilities, disclaiming inspection, and recommending specialist evaluation if the client wants information about the installation and condition of these facilities.

HISOP Section 11 - Insulation and Ventilation

INTRODUCTION

11.1.A.1 – INSULATION AND VAPOR RETARDER INSPECTION REQUIREMENTS

The objective of the insulation and vapor retarder inspection is to identify and to report visual evidence that these components may not be functioning properly, or that they may be significantly deficient. These components rarely create an unsafe condition, but if one exists, the inspector should report the unsafe condition, usually as an implication of the improper functioning or significant deficiency that is causing the unsafe condition (see also Clause 2.2.1.B.1, unsafe). The service life of these components, if properly installed and maintained, should approximate the life of the building. End of service life is based on component condition more than on age.

Inspection of insulation and vapor retarders is to determine the presence and condition of visible components between conditioned and unconditioned spaces. Inspectors are not required to: measure insulation, determine the R-value of insulation, determine if the insulation is adequate, determine if the insulation complies with current or past requirements, and determine if vapor retarders are adequate, or if they comply with current or past requirements (see also Clauses 13.1.B.1, 13.2.A.3, and 13.2.A.8). Some inspectors provide some of these services, but this is not required.

Inadequate insulation and improper vapor retarder installation can be the cause of significant deficiencies, such as condensation on components, which can cause component deterioration. Inspectors should report visible evidence of significant deficiencies that might be caused by insulation and vapor retarders, and should recommend evaluation to determine the cause and recommended actions (see also Clauses 2.2.B.1 and 13.2.A.3). Note that air leaks between conditioned and unconditioned spaces can be a more important cause of condensation deficiencies than insulation and vapor retarders.

Some insulation is known to be problematic. Vermiculite insulation often contains asbestos. Urea-formaldehyde foam insulation installed during the 1970s and early 1980s had off-gassing problems, but this is no longer a concern. Inspectors are not required to report the presence of problematic components (see also Clauses 13.2.A.12 and 13.2.A.17). Most inspectors report the presence of visible vermiculite and recommend evaluation to determine if it contains asbestos. Urea-formaldehyde foam was used mostly in wall cavities, so it is rarely visible during a home inspection.

Inspection of radiant barriers is out-of-scope because they are not specifically addressed in the HISOP (see also Clause 13.1.A). Inspectors should report installation of a radiant barrier that creates an improper vapor retarder, or that creates a potentially unsafe condition because of contact with electrical cables. Inspectors may consider reporting the presence of radiant barriers, disclaiming inspection, and recommending specialist evaluation if the client wants information about the installation and condition of the radiant barrier.

Insulation and Vapor Retarders Not Functioning Properly

The primary function of insulation is to slow the flow of heat between conditioned and unconditioned spaces. The primary function of vapor retarders is to slow the flow of moisture between conditioned and unconditioned spaces. It is not possible to determine if insulation and vapor retarders are functioning properly without measuring heat and water vapor flow. Insulation and vapor retarder defects are, therefore, significant deficiencies.

Significantly Deficient Insulation and Vapor Retarders

Some common examples of insulation and vapor retarder significant deficiencies include: absent insulation or vapor retarder where required, significantly disturbed, compressed, or dirty insulation, insulation not cut to fit location where installed, insulation not in contact with an air barrier, vapor retarder installed in the wrong direction, insulation or vapor retarder in contact with heat sources such as vents, multiple installed vapor retarders, insulation or vapor retarders blocking ventilation openings, and exposed Kraft paper vapor retarder.

11.1.A.2 – ATTIC AND CRAWLSPACE VENTILATION INSPECTION REQUIREMENTS

The objective of the attic and crawlspace ventilation inspection is to identify and report visual evidence that these components may not be functioning properly, or that they may be significantly deficient. These components rarely create an unsafe condition, but if one exists, the inspector should report the unsafe condition, usually as an implication of the improper functioning or significant deficiency that is causing the unsafe condition (see also Clause 2.2.1.B.1, unsafe). The service life of these components, if properly installed and maintained, should approximate the life of the building. End of service life is based on component condition more than on age.

Inspection of attic and crawlspace ventilation components is to determine the presence and condition of visible ventilation components. Inspectors are not required to: measure ventilation openings, determine if the ventilation is adequate, determine if ventilation complies with current or past requirements, or to determine if vapor retarders installed as part of the ventilation system are adequate, or if they comply with current or past requirements (see also Clauses 13.1.B.1, 13.2.A.3, and 13.2.A.8). Some inspectors provide some of these services, but this is not required.

Some attics are equipped with powered ventilation fans that activate at a set temperature. Inspectors are not required to test these fans, or to report if they are not operating during the inspection (see also Clauses 13.1.B.2.a and 13.2.A.1). Inspectors may consider reporting the presence of these fans, disclaiming inspection, and recommending specialist evaluation if the client wants information about the installation and condition of the fans.

Some attics and crawlspaces are designed to be unventilated. Specifications for unventilated attics and crawlspaces depend on the climate zone where the building is located. Inspectors are not required to determine if the unventilated attic or crawlspace complies with current or past requirements (see also Clauses 13.1.B.1, 13.2.A.3, and 13.2.A.8). Inspectors may consider disclaiming inspection of unventilated attics and crawlspaces, and recommending evaluation by a qualified specialist if the client wants assurance about the condition of these areas.

Attic and Crawlspace Ventilation Not Functioning Properly

The primary function of attic ventilation is to remove excess heat and moisture from the attic in the summer, to remove excess moisture from the attic in the winter, and to maintain attic temperature near the outdoor temperature in the winter.

The stated function of crawlspace ventilation is to remove excess moisture from the crawlspace. There is increasing skepticism about whether crawlspace ventilation actually performs this function, especially in warm, humid climate zones; however, crawlspace ventilation, or an approved unventilated crawlspace, is required for all crawlspaces.

Some common examples of attic ventilation not functioning properly include: evidence of condensation such as rusted nails, widespread dark stains on framing components, excessive fungal growth on wood, and deterioration of framing components. Note that these indications can have other causes such as water leaks and air leaks from conditioned spaces below the attic.

Some common examples of crawlspace ventilation not functioning properly include: excessive fungal growth on wood, deteriorated wood, deteriorated insulation, damp insulation, and excessive condensation on air conditioning ducts, especially metal ducts. Note that these indications can have other causes.

Significantly Deficient Attic and Crawlspace Ventilation

Some common examples of attic and crawlspace ventilation significant deficiencies include: deteriorated, damaged, or inoperative openings or opening covers, openings painted shut, openings blocked by insulation, and no eave or ridge ventilation openings. Inspectors are not required to operate operable opening covers.

11.1.A.3 AND 11.1.A.4 – EXHAUST SYSTEM INSPECTION REQUIREMENTS

The objective of the exhaust system inspection is to identify and to report visual evidence that these systems may not be functioning properly, or that they may be significantly deficient. These systems rarely create an unsafe condition, but if one exists, the inspector should report the unsafe condition, usually as an implication of the improper functioning or significant deficiency that is causing the unsafe condition (see also Clause 2.2.1.B.1, unsafe). The service life of exhaust fans is limited; however, end of service life is based on component condition more than on age. The service life of exhaust ducts, if properly installed and maintained, should approximate the life of the building. End of service life is based on component condition more than on age.

Inspection of exhaust systems is to determine the presence and condition of visible system components. Inspectors are not required to measure exhaust duct length or opening size; and are not required to determine if the exhaust system air flow rate is adequate, or if it complies with current or past requirements (see also Clauses 13.1.B.1, 13.2.A.3, and 13.2.A.8). Some inspectors provide some of these services, but this is not required.

For many years, an exhaust system with a duct terminating outdoors has been required in a bathroom and in a kitchen without an operable window. An exhaust system is recommended in all bathrooms and kitchens. Most kitchens have an operable window, so an exhaust system with a duct terminating outdoors is not required in most kitchens. A recirculating fan is not required in a kitchen.

Inspection of heat recovery ventilation systems, energy recovery ventilation systems, and other central exhaust and ventilation systems is out-of-scope (see also Clause 8.2.A.6).

Exhaust Systems Not Functioning Properly

The primary function of exhaust systems is to remove moisture and odors from the building. Some common examples of exhaust systems not functioning properly include: fan does not activate, duct does not terminate outside of the building, duct termination is blocked or damper is stuck shut.

Significantly Deficient Exhaust Systems

Some common examples of exhaust systems significant deficiencies include: deteriorated, damaged, or disconnected duct, duct terminates too close to a building opening, clothes dryer duct is not four inches in diameter, lint at clothes dryer duct termination, no screen on exhaust duct termination (except clothes dryer), and screen on clothes dryer exhaust termination.

11.1.B.1 – INSULATION AND VAPOR RETARDER DESCRIPTIONS

This clause requires the inspector to describe the types of visible insulation and vapor retarders installed in the building. As previously discussed in Clause 2.2.B.1, a description should allow the reader to distinguish the described components from similar components. If multiple types of insulation and vapor retarders are present, the inspector should describe each type of insulation and vapor retarder. Describing Class III vapor retarders, such as paint, is not required. Describing concealed insulation and vapor retarders, such as in wall cavities or behind air barriers, is not required (see also Clauses 13.1.B.2.a, and 13.2.A.1).

Typical Insulation and Vapor Retarder Descriptions

The following descriptions are examples of insulation and vapor retarder descriptions, and are not intended to be a complete list of all such descriptions. The attic floor insulation is loose-fill cellulose. The attic floor insulation is fiberglass batts with a Kraft paper vapor retarder. The crawlspace floor joist insulation is fiberglass batts without a vapor retarder.

11.1.B.2 – INSULATION ABSENCE DESCRIPTION

This clause emphasizes that inspectors are required to report if there is no insulation between conditioned and unconditioned space. Absence of insulation is a significant deficiency, and inspectors should report this condition as such.

11.2 – INSULATION DISTURBANCE INSPECTION LIMITATION

Inspectors are not required to move or to disturb insulation to perform any inspection procedure required by the American Society of Home Inspectors (ASHI) Standard of Practice for Home Inspections (HISOP) (see also Clauses 13.1.B.2.a, and 13.2.A.1).

HISOP Section 12 - Fireplaces and Fuel-Burning Appliances

INTRODUCTION

This section contains the requirements and specific limitations for inspecting fireplaces and fuel-burning appliances. This section includes masonry fireplaces, factory-built wood-burning fireplaces, solid-fuel-burning stoves, and solid-fuel-burning fireplace inserts. This section also includes fuel-gas and liquid-fuel-burning appliances such as vented and un-vented gas fireplaces, gas logs, gas fireplace inserts, and gas fire-starters. Vent systems and chimneys serving in-scope systems are also included.

12.1.A.1, 12.1.A.2, AND 12.1.A.3 – FIREPLACE, FUEL-BURNING APPLIANCE, CHIMNEY, AND VENT SYSTEM INSPECTION REQUIREMENTS

The objective of the fireplace, fuel-burning appliance, chimney and vent system inspection is to identify and to report visual evidence that these systems may not be functioning properly, or that they may be significantly deficient. These components can create an unsafe condition, and if one exists, the inspector should report the unsafe condition, usually as an implication of the improper functioning or significant deficiency that is causing the unsafe condition (see also Clause 2.2.1.B.1, unsafe). The service life of masonry fireplaces and masonry chimneys, if properly installed and maintained, should approximate the life of the building. Other systems have a limited service life; however, end of service life is based on component condition more than on age.

Inspectors are not required to, and in most cases should not, test or operate fireplaces and fuel-burning appliances unless operation is by a normal operating control, such as a wall switch. Inspectors are not required to, and in most cases should not, use a flame source, such as a lighter or a match, to ignite any fuel or pilot light (see also Clauses 13.2.C.2 and 13.2.F.6).

Fireplace, Fuel-Burning Appliance, Chimney, and Vent System Not Functioning Properly

The primary functions of these systems are to safely burn fuel, and to conduct combustion products outside of the building. Some common examples of these systems not functioning properly include: soot stains around the fireplace, soot stains around the chimney or vent, soot stains at the chimney or vent termination, moisture stains, deterioration, or other evidence of condensation or water leakage in the fireplace, chimney, or vent.

<u>Significantly Deficient Fireplace, Fuel-Burning Appliance, Chimney, and Vent System</u>

Some common examples of fireplace, fuel-burning appliance, chimney, and vent system significant deficiencies include: chimney leaning or not plumb, inadequate distance to combustible materials, deterioration or damage, large cracks in bricks or mortar, creosote buildup, lintel bowed or significantly rusted, damper damaged or not operational, inadequate chimney height, absent chimney cricket, and gaps between the fireplace and chimney.

12.2.A.1 – FLUE INTERIOR INSPECTION LIMITATION

Refer to Clause 5.2.B for the discussion of this limitation.

12.2.A.2 – SCREENS AND DOORS INSPECTION LIMITATION

Inspectors are not required to inspect fireplace screens, doors, and fireplace accessories such as tools and grates (see also Clauses 13.2.E.2, and 13.2.E.3).

12.2.A.3 – SEALS AND GASKETS INSPECTION LIMITATION

Many fuel-burning appliances, such as fuel-burning stoves and fuel-burning fireplace inserts, have seals and gaskets installed at various locations in the system. Inspection of visible and readily accessible seals and gaskets, such as those around the door, is in-scope. Inspection of seals and gaskets that are not visible and readily accessible is out-of-scope (see also Clauses 13.2.B.2.a, 13.2.A.1, and 13.2.F.4).

12.2.A.4 – AUTOMATIC FUEL FEED INSPECTION LIMITATION

Some solid-fuel-burning appliances, such as pellet stoves, are equipped with a device that continually supplies the appliance with fuel. It can be difficult to determine if these devices are functioning properly during a home inspection. Inspectors are not required to test or to report about these devices.

HISOP Section 13 - General Limitations and Exclusions

INTRODUCTION

This section contains limitations and exclusions that apply to all HISOP sections. References to these limitations and exclusions are included with the discussion of many clauses of the American Society of Home Inspectors (ASHI) Standard of Practice for Home Inspections (HISOP); however, the reader should not assume that lack of a reference means that a limitation or exclusion does not apply.

13.1.A, AND 13.2.F.2 – GENERAL HISOP SCOPE LIMITATION

A building and the land on which it sits consists of hundreds of systems and components. Each system and component could have hundreds of types of deficiencies. It is not practical or cost-effective for a home inspection to include all possible systems, components, and deficiencies.

A home inspection includes only those systems, components, and deficiencies that are specifically stated in the HISOP, and that are located in areas that are required to be inspected (see also Clauses 3.2.C, 3.2.D, and 13.2.A.1). All other systems, components, and deficiencies are out-of-scope based on these clauses.

13.1.B.1 – TECHNICALLY EXHAUSTIVE LIMITATION

Technically exhaustive is a defined term that excludes from scope procedures that go beyond the limited visual inspection and limited appliance operation requirements contained in the definitions of: home inspection, inspect, dismantling, and normal operating controls. For example, it is out-of-scope to use a tape measure to measure a floor joist span, then use that measurement to determine if the joist is over-spanned (see also Clauses 13.2.A.3 and 13.2.A.8). For example, it is out-of-scope to use instruments such as combustible gas detectors, moisture meters, circuit analyzers, infrared cameras, drones, and carbon monoxide detectors. For example, it is out-of-scope, and improper, for a home inspector to provide services such as engineering, architecture, contracting, or other services for which a license is usually required even if the inspector is licensed to do so (see also Clauses 13.2.A.3, 13.2.B.1, 13.2.B.2, and 13.2.B.3). Home inspectors should not report conclusions or make recommendations that could be interpreted as providing these services.

13.1.B.2.A – CONCEALED AND LATENT CONDITIONS AND CONSEQUENTIAL DAMAGES LIMITATION

A home inspection is a limited visual inspection. Deficiencies that are concealed, or that are located in areas that are not readily accessible, are out-of-scope (see also Clause 13.2.A.1). A concealed defect may occur because of intentional concealment, such as painting a water stain. A concealed defect may occur because the defect is behind occupant belongings, such as water damage behind a piece of furniture.

Latent defects are out-of-scope. A latent defect is one that is not discoverable by in-scope visual inspection procedures. A latent defect may occur only under certain conditions, such as poor draft in a flue that occurs only during cold weather, or that occurs only when the wind blows from a certain direction.

Consequential damages are out-of-scope. Consequential damages are damages that occur as a result of a separate and different defect. For example, significantly deficient flashing may allow water infiltration into the building, which may provide moisture for fungal growth, which may cause an allergic reaction in some people. The defect in this case is the flashing. The consequential damages are the fungal growth and the allergic reaction.

13.1.B.2.B – COSMETIC IMPERFECTIONS LIMITATION

Cosmetic imperfections are out-of-scope. Cosmetic imperfections are those that do not significantly affect the performance of the intended function of a component. Examples include imperfections in paint and stain, typical damage and deterioration of visible surfaces, and normal wear and tear. Home inspections provide more benefits to clients when the inspector focuses on functional deficiencies, and focuses on areas where people usually do not go, such as attics and crawlspaces.

13.1.C – BUILDINGS TO WHICH THE HISOP APPLIES

The HISOP is intended to be used to inspect completed buildings that contain four or fewer residential dwelling units, and the attached and detached garages and carports. Other types of buildings, and other types of occupancies, such as apartments, commercial, and industrial, may be constructed using different materials, construction methods, and standards; therefore, the HISOP does not fully apply to other types of buildings and occupancies. The HISOP is also not intended to be used to inspect buildings under construction. Inspectors should use the ASHI Standard of Professional Practice for Residential Predrywall Inspections for such inspections.

Inspectors should use caution when inspecting buildings that are not within the intended scope of the HISOP. Inspectors should also use caution when inspecting out-of-scope buildings because such inspections may require additional training and experience. Appropriate warnings and disclaimers are prudent in order to help ensure against misunderstandings with clients if the inspector elects to perform inspections on buildings that are not within the scope of the HISOP.

13.1.D – RELATIONSHIP OF THE HISOP TO STATE STANDARDS OF PRACTICE

The state-mandated standard of practice (SOP) is the applicable SOP in states that require home inspectors to obtain a state-issued license, and that have a state-mandated SOP. Where the state-mandated SOP and the HISOP conflict, this clause clarifies that inspectors should comply with the state-mandated SOP. For example, the state-mandated SOP may require inspection of additional systems and components. Where the state-mandated SOP does not address a system or component, inspectors should comply with the HISOP.

13.1.E – REDUNDANCY

Some clauses in the HISOP use different wording to address the same or similar concepts. This clause clarifies that this redundancy is an attempt to clarify and to reinforce the concept.

13.2.A.1 – READILY ACCESSIBLE LIMITATION

Refer to the discussion of readily accessible in Clause 2.2.A for the discussion of this limitation.

13.2.A.2, AND 13.2.A.6 – REMAINING SERVICE LIFE LIMITATION

Unless a system or component is not functioning and cannot be repaired (the remaining service life is zero), it is not possible to forecast the remaining service life of a system or component, or to forecast when and how the system or component may fail. Inspectors are not required to, and should not, report about or speculate about, the remaining service life of a system or a component.

It is important to note that reporting about systems and components that may be near the end of their service life is different from speculating about the remaining service life. Refer to Clause 2.2.B.1 for more discussion about this topic.

13.2.A.3 – STRENGTH, ADEQUACY, EFFECTIVENESS, AND EFFICIENCY LIMITATION

Inspectors are not required to, and in most cases should not, determine and report about these properties or characteristics. Doing so is technically exhaustive, and usually requires expertise and procedures that are out-of-scope for home inspectors and home inspections (see also Clauses 13.1.B.1, 13.2.B.2, and 13.2.B.3).

13.2.A.4 – CAUSE OF DEFICIENCY LIMITATION

Inspectors are not required to, and in most cases should not, determine or speculate about why a system or component is deficient. Determining the cause of a deficiency often requires technically exhaustive procedures, and may require skills and equipment that the inspector does not possess. Inspectors should report the deficiency and recommend evaluation to determine the cause of the deficiency (see also Clauses 13.1.B.1, 13.2.B.2, and 13.2.B.3).

13.2.A.5 – METHODS AND COST TO CURE LIMITATION

Inspectors are not required to, and in most cases should not, specify or design the methods that might be used to correct a deficiency, and should not provide estimates about the cost to correct the deficiency. Inspectors often do not have complete information about the causes and the extent of the deficiency, or about alternatives that might be available to correct the deficiency. Specifying or designing repairs may be construed as practicing a licensed trade or profession.

Note that inspectors may offer these services, if the inspector is qualified and willing to do so (see also Clause 2.3.B), and that these services are common practice in some markets. Note that inspectors may not repair, replace, or upgrade for compensation systems and components specified in the HISOP for one year after the inspection (see also Code of Ethics 1.F).

13.2.A.7, 13.2.A.9, AND 13.2.A.10 – USE AND VALUE OF PROPERTY LIMITATIONS

Inspectors are not required to, and should not, make statements that could be construed as dealing with issues surrounding property use and zoning, property market value, and whether the property should be purchased. Inspectors usually do not have the necessary information and expertise to make such statements. Inspectors should defer these issues to the appropriate professionals, such as appraisers, real estate attorneys, and real estate agents.

13.2.A.8 – CODES AND MANUFACTURER'S INSTRUCTIONS LIMITATION

Inspectors are not required to, and should not, make statements that could be construed as stating that a deficiency or condition violates past or present government laws and regulations, such as building codes or zoning ordinances. Only people authorized by the appropriate government agency may make such determinations.

Inspectors are not required to determine if a system or component complies with building codes, zoning ordinances, manufacturer's installation instructions, industry guidelines, and similar rules and guidelines. Installation of many systems and components is based on these rules and guidelines, and these rules and guidelines often vary by manufacturer and model. It is not possible for inspectors to know, or to determine, all such rules and guidelines. Inspectors may consult and use rules and guidelines to identify possible deficiencies, but this is not required.

13.2.A.11, 13.2.A.12, AND, 13.2.A.13 – HAZARDOUS OR HARMFUL PLANTS, ANIMALS, AND SUBSTANCES LIMITATIONS

Inspectors are not required to, and should not, make statements that could be construed as stating conclusions about the presence or absence of environmental risks and hazardous or harmful plants, animals, or substances, unless the inspector is qualified and licensed (if required) to make such statements, and unless the inspector has performed the necessary inspections and tests. This includes, but is not limited to, statements about wood destroying organisms, pests and vermin, fungi, allergens, toxins, carcinogens, electromagnetic radiation, noise, radon and other radioactive substances, and other contaminates.

Inspectors are not required to, and should not, make statements that could be construed as stating conclusions about systems and methods used to control, mitigate, or remove hazardous plants, animals, and substances, unless the inspector is qualified and licensed (if required) to make such statements, and unless the inspector has performed the necessary inspections and tests.

13.2.A.14 – OPERATING COST LIMITATION

Inspectors are not required to, and should not, make statements that could be construed as dealing with such issues as the cost of operating systems and components, and about the cost savings that might be realized by replacing or upgrading systems and components. Inspectors usually do not have the necessary information and expertise to make such statements. Inspectors should defer these issues to the appropriate professionals.

13.2.A.15 – ACOUSTICAL PROPERTIES LIMITATION

Inspectors are not required to, and should not, make statements that could be construed as dealing with issues related to sound transmission through building components, about the presence or absence of sound or noise in and around the property, and about methods and costs of reducing sound transmission through building components. Inspectors should defer these issues to the appropriate professionals.

13.2.A.16 – SOIL CONDITIONS LIMITATION

Inspectors are not required to, and should not, make statements that could be construed as dealing with issues related to the presence or absence of unstable soils, sliding soils, sinkholes, and other similar issues, or about methods and costs of dealing with these issues. Inspectors should defer these issues to the appropriate professionals, such as geotechnical engineers.

13.2.A.17 – RECALL AND PRODUCT LIABILITY LIMITATION

Inspectors are not required to report whether systems or components are or have been subject to any type of controversy, such as recalls, litigation, and settlements. Inspectors may provide this service, but if doing so, appropriate limitations and disclaimers are prudent.

13.2.B.1, 13.2.B.2, AND, 13.2.B.3 – ADHERENCE TO LAWS AND REGULATIONS

Inspectors are not required to, and should not, perform acts or services that may violate government laws or regulations. Examples include performing services or making statements that could be construed as practicing trades or professions, such as engineering, architecture, and contracting, that often require a separate government license.

Inspectors may provide additional services if the inspector is qualified and licensed (if required) to do so, and if the inspector has performed the necessary inspections, data gathering, and tests (see also Clause 2.3.B). Inspectors who provide additional services should obtain a written agreement from the client to do so. The agreement should include the scope and limitations of the additional services, and the cost of those services.

13.2.B.4 – WARRANTIES AND GUARANTEES LIMITATION

Inspectors are not required offer or to provide warranties or guarantees of any kind covering their services, or covering systems and components in the inspected building. Inspectors may provide this service, but if doing so, appropriate limitations and disclaimers are prudent.

13.2.C.1 AND 13.2.C.2 – SHUT DOWN SYSTEMS LIMITATION

Inspectors are not required to, and in most cases should not, test or operate systems and components that are shut down, or that do not respond to normal operating controls. Examples of these situations include turning on water, electricity, or gas that is turned off at the meter, at the house or at the appliance shutoff valve, at the fixture stop valve, at the service equipment, or at the overcurrent device. Inspectors may perform these procedures, but caution is prudent, and inspectors should return the system or component to the state it was in prior to the inspection.

13.2.C.3 – SHUTOFF AND STOP VALVE LIMITATION

Inspectors are not required to, and in most cases should not, test or operate shutoff valves and stop valves. Examples of these valves include the house water service valve, water heater cold water shutoff valve, and stop valves at toilets and sinks. Inspectors may perform these procedures, but caution is prudent, and inspectors should return the valve to the state it was in prior to the inspection.

13.2.C.4 – AUTOMATIC SAFETY CONTROLS LIMITATION

Inspectors are not required to, and in most cases should not, test or operate automatic safety controls. Examples of these controls include auxiliary condensate cutoff switches at evaporator coils and at pans under evaporator coils, thermocouples at gas valves, flame sensors in furnaces, and overcurrent protection devices such as circuit breakers.

13.2.D.1 AND 13.2.F.1 – DANGER TO PEOPLE OR PROPERTY LIMITATION

Inspectors are not required to, and should not, enter areas or perform procedures that are likely to expose people (including the inspector) or the property to the risk of injury or damage. The inspector is solely responsible for assessing the risk, and for determining whether it is prudent to assume the risk. The inspector's assessment is based on the conditions at the time of the inspection. Reasonable inspectors can reach different conclusions about whether or not it is prudent to assume the risk (see also Clause 13.2.E.4).

Inspectors who elect not to enter an area or perform a procedure required by the HISOP must report this decision in the inspection report, and must explain that the decision was based on the risk of injury or the risk of property damage (see also Clause 2.2.B.4). It is highly recommended, but not required, that inspectors also include a recommendation that the system or component be inspected, and include a warning that deficiencies could exist.

13.2.D.2 – ATTIC AND CRAWLSPACE ENTRY LIMITATION

Inspectors are not required to enter attic and crawlspace areas that are not readily accessible. This limitation is discussed in Clauses 2.2.A, 3.2.C, 3.2.D, 13.2.D.1, and 13.2.F.1.

13.2.E.1 – UNDERGROUND COMPONENT INSPECTION LIMITATION

Inspectors are not required to inspect components located underground. This limitation specifically mentions underground fuel storage tanks, such as those for oil and propane, and it includes components that could indicate the presence of underground fuel storage tanks, such as vents, fill tubes, stains, and dead vegetation (see also Clause 6.1.A.6). Inspectors should consider reporting obvious visible indications of the presence of these components, disclaiming inspection, and recommending specialist evaluation if the client wants information about the installation and condition of these components.

This limitation includes all underground components. Other examples of underground components include the water service pipe, building sewer pipe, and electrical service lateral (see also Clauses 13.1.B.2.a and 13.2.A.1).

13.2.E.2 – ITEMS NOT INSTALLED LIMITATION

Inspectors are not required to, and in most cases should not, inspect systems, components, and other items that are not installed. Installed is a defined term. These items are usually considered personal property that usually does not convey with the building. These items could be exchanged for other items that the inspector did not inspect, thus creating the potential for misunderstanding with the client.

13.2.E.3 – DECORATIVE ITEMS LIMITATION

Inspectors are not required to inspect decorative items (see also Clauses 4.2.A, 10.2.A, 10.2.B, 10.2.C, and 13.1.B.2.b). In addition to the items referenced in these clauses, decorative items could include mirrors, medallions, murals, or other permanent artwork.

13.2.E.5 – DETACHED STRUCTURES LIMITATION

Inspectors are not required to inspect detached structures other than garages and carports. Refer to Clause 4.2.E for the discussion of this limitation.

13.2.E.6 – COMMON ELEMENTS AND AREAS LIMITATION

Inspectors are not required to, and in most cases should not, inspect the common systems, components, and areas of multi-family buildings. A multi-family building is one that has three or more residential dwelling units in one building, or one that has joint ownership of common systems, components, and areas outside of the dwelling unit walls. These buildings are sometimes referred to as condominiums, but this term describes a form of real property ownership, not a type of building.

These common systems, components, and areas may include, but are not limited to, landscaping plants, driveways, parking areas, walkways, exterior stairways, guards, handrails, structural components, exterior wall coverings and flashing, roof coverings and flashing, and some components of the plumbing, HVAC, and electrical systems. The HVAC system and the electrical panel inside the unit are usually not considered to be common systems and components.

Inspectors may consider reporting clearly visible safety concerns and clearly visible damage or deterioration in common systems and components, and in areas that could result in a special assessment to repair or to replace the systems, components, or other items, but this is not required.

13.2.E.7 – MULTIPLE SIMILAR COMPONENTS LIMITATION

Inspectors are not required to inspect every inch of every in-scope system or component. Inspectors seek to balance the time required to perform an inspection against reasonable compensation for their time and expertise. Inspectors must, therefore, allocate time based on the inspector's training and experience to determine the of types of deficiencies that are most likely to occur, and where these deficiencies might be located. This means inspecting similar components on a random basis, assuming that if the inspected components are in acceptable condition, similar components are probably also in acceptable condition.

Inspectors should consider including this and other similar limitations in the inspection agreement, and should consider discussing this and other similar limitations with the client before the inspection.

13.2.E.8 – OUTDOOR COOKING APPLIANCE LIMITATION

Inspectors are not required to inspect outdoor cooking appliances. This includes all components of outdoor kitchens, whether installed or portable. Examples of out-of-scope appliances include grills, barbecues, ovens, sinks, and refrigerators (see also Clauses 6.1.A.1, 6.1.A.2, and 10.2.G). Inspectors may provide this service, but if doing so, appropriate limitations and disclaimers are prudent.

13.2.F.3 – MOVING MATERIALS LIMITATION

Inspectors are not required to, and in most cases should not, move personal property, equipment, appliances, soil, snow and ice, debris, or any other material. Refer to the discussion of readily accessible in Clause 2.2.A for further discussion of this limitation.

13.2.F.4 – DISMANTLING LIMITATION

Inspectors are not required to, and in most cases should not, dismantle systems and components. Refer to the discussion of dismantle in Clause 2.2.A for further discussion of this limitation.

13.2.F.5 – RESETTING CLOCKS AND TIMERS LIMITATION

Inspection procedures that are required or recommended in the HISOP can cause temporary disruption of electrical power to clocks, timers, computers, routers, video recording devices, and other devices. Inspectors should take reasonable precautions to avoid such disruption, unless necessary to perform the inspection; however, sellers should have a reasonable expectation of minor inconvenience as a result of the home inspection. Inspectors are not required to reset, reprogram, or adjust devices that are disrupted during the inspection.

13.2.F.6 – IGNITING FIRES LIMITATION

Inspectors are not required to, and in most cases should not, ignite or extinguish any flame that requires manual ignition. This includes appliances such as gas cooking appliances of all types, furnaces and room heaters, liquid, gas, and solid-fuel burning fireplaces, stoves, and fireplace inserts. Inspectors increase liability risk when performing this procedure.

13.2.F.7 – PROBING LIMITATION

Refer to the discussion of probing in Clause 2.2.A for further discussion of this limitation.

ASHI Standard of Practice for Home Inspections

1. INTRODUCTION

The American Society of Home Inspectors®, Inc. (ASHI®) is a not-for-profit professional society established in 1976. Membership in ASHI is voluntary and its members are private home *inspectors*. ASHI's objectives include promotion of excellence within the profession and continual improvement of its members' *inspection* services to the public.

2. PURPOSE AND SCOPE

2.1 The purpose of this document is to establish a minimum standard (Standard) for *home inspections* performed by *home inspectors* who subscribe to this Standard. *Home inspections* performed using this Standard are intended to provide the client with information about the condition of *inspected systems* and *components* at the time of the *home inspection*.

2.2 The *inspector* shall:

A. *inspect readily accessible*, visually observable, *installed systems* and *components* listed in this Standard.

B. provide the client with a written report, using a format and medium selected by the *inspector*, that states:

1. those *systems* and *components inspected* that, in the professional judgment of the *inspector*, are not functioning properly, significantly deficient, *unsafe*, or are near the end of their service lives,

2. recommendations to correct, or monitor for future correction, the deficiencies reported in 2.2.B.1, or items needing *further evaluation* (Per Exclusion 13.2.A.5 the *inspector* is NOT required to determine methods, materials, or costs of corrections.),

3. reasoning or explanation as to the nature of the deficiencies reported in 2.2.B.1, that are not self-evident,

4. those *systems* and *components* designated for *inspection* in this Standard that were present at the time of the *home inspection* but were not *inspected* and the reason(s) they were not *inspected*.

C. adhere to the ASHI® Code of Ethics for the Home Inspection Profession.

2.3 This Standard is not intended to limit the *inspector* from:

A. including other services or *systems* and *components* in addition to those required in Section 2.2.A.

B. designing or specifying repairs, provided the *inspector* is appropriately qualified and willing to do so.

C. excluding *systems* and *components* from the inspection if requested or agreed to by the client.

3. STRUCTURAL COMPONENTS

3.1 The *inspector* shall:

A. *inspect structural components* including the foundation and framing.

B. *describe*:

1. the methods used to *inspect under-floor crawlspaces* and attics.

2. the foundation.

3. the floor structure.

4. the wall structure.

5. the ceiling structure.

6. the roof structure.

3.2 The *inspector* is NOT required to:

A. provide *engineering* or architectural services or analysis.

B. offer an opinion about the adequacy of structural *systems* and *components*.

C. enter *under-floor crawlspace* areas that have less than 24 inches of vertical clearance between *components* and the ground or that have an access opening smaller than 16 inches by 24 inches.

D. traverse attic load-bearing *components* that are concealed by insulation or by other materials.

4. EXTERIOR

4.1 The *inspector* shall:

A. *inspect*:

1. *wall coverings*, flashing, and trim.

2. exterior doors.

3. attached and adjacent decks, balconies, stoops, steps, porches, and their associated railings.

4. eaves, soffits, and fascias where accessible from the ground level.

5. vegetation, grading, surface drainage, and retaining walls that are likely to adversely affect the building.

6. adjacent and entryway walkways, patios, and driveways.

B. *describe wall coverings.*

4.2 The *inspector* **is NOT required to** *inspect*:

A. screening, shutters, awnings, and similar seasonal accessories.

B. fences, boundary walls, and similar structures.

C. geological and soil conditions.

D. *recreational facilities.*

E. outbuildings other than garages and carports.

F. seawalls, break-walls, and docks.

G. erosion control and earth stabilization measures.

5. ROOFING

5.1 The *inspector* **shall:**

A. *inspect*:

1. roofing materials.

2. *roof drainage systems.*

3. flashing.

4. skylights, chimneys, and roof penetrations.

B. *describe*:

1. roofing materials.

2. methods used to *inspect* the roofing.

5.2 The inspector is NOT required to *inspect*:

A. antennas.

B. interiors of vent *systems*, flues, and chimneys that are not *readily accessible.*

C. other *installed* accessories.

6. PLUMBING

6.1 The *inspector* shall:

A. *inspect*:

1. interior water supply and distribution *systems* including fixtures and faucets.

2. interior drain, waste, and vent *systems* including fixtures.

3. water heating equipment and hot water supply *systems*.

4. vent *systems*, flues, and chimneys.

5. fuel storage and fuel distribution *systems*.

6. sewage ejectors, sump pumps, and related piping.

B. *describe*:

1. interior water supply, drain, waste, and vent piping materials.

2. water heating equipment including energy source(s).

3. location of main water and fuel shut-off valves.

6.2 The *inspector* is NOT required to:

A. *inspect*:

1. clothes washing machine connections.

2. interiors of vent *systems*, flues, and chimneys that are not *readily accessible*.

3. wells, well pumps, and water storage related equipment.

4. water conditioning *systems*.

5. solar, geothermal, and other renewable energy water heating *systems*.

6. manual and automatic fire extinguishing and sprinkler *systems* and landscape irrigation *systems*.

7. septic and other sewage disposal *systems*.

B. determine:

1. whether water supply and sewage disposal are public or private.

2. water quality.

3. the adequacy of combustion air *components*.

C. measure water supply flow and pressure, and well water quantity.

D. fill shower pans and fixtures to test for leaks.

7. ELECTRICAL

7.1 The *inspector* shall:

  A. *inspect*:

   1. service drop.

   2. service entrance conductors, cables, and raceways.

   3. service equipment and main disconnects.

   4. service grounding.

   5. interior *components* of service panels and subpanels.

   6. conductors.

   7. overcurrent protection devices.

   8. a *representative number* of *installed* lighting fixtures, switches, and receptacles.

   9. ground fault circuit interrupters and arc fault circuit interrupters.

  B. *describe*:

   1. amperage rating of the service.

   2. location of main disconnect(s) and subpanels.

   3. presence or absence of smoke alarms and carbon monoxide alarms.

   4. the predominant branch circuit *wiring method*.

7.2 The *inspector* is NOT required to:

  A. *inspect*:

   1. remote control devices.

   2. or test smoke and carbon monoxide alarms, security *systems*, and other signaling and warning devices.

   3. low voltage wiring *systems* and *components*.

   4. ancillary wiring *systems* and *components* not a part of the primary electrical power distribution system.

   5. solar, geothermal, wind, and other renewable energy *systems*.

  B. measure amperage, voltage, and impedance.

  C. determine the age and type of smoke alarms and carbon monoxide alarms.

8. HEATING

8.1 The *inspector* shall:

  A. open *readily openable access panels*.

  B. *inspect*:

    1. *installed* heating equipment.

    2. vent *systems*, flues, and chimneys.

    3. distribution *systems*.

  C. *describe*:

    1. energy source(s).

    2. heating *systems*.

8.2 The *inspector* is NOT required to:

  A. *inspect*:

    1. interiors of vent *systems*, flues, and chimneys that are not *readily accessible*.

    2. heat exchangers.

    3. humidifiers and dehumidifiers.

    4. electric air cleaning and sanitizing devices.

    5. heating *systems* using ground-source, water-source, solar, and renewable energy technologies.

    6. heat-recovery and similar whole-house mechanical ventilation *systems*.

  B. determine:

    1. heat supply adequacy and distribution balance.

    2. the adequacy of combustion air *components*.

9. AIR CONDITIONING

9.1 The *inspector* shall:

  A. open *readily openable access panels*.

  B. *inspect*:

    1. central and permanently *installed* cooling equipment.

    2. distribution *systems*.

  C. *describe*:

    1. energy source(s).

    2. cooling *systems*.

9.2 The *inspector* is NOT required to:

A. *inspect* electric air cleaning and sanitizing devices.

B. determine cooling supply adequacy and distribution balance.

C. *inspect* cooling units that are not permanently *installed* or that are *installed* in windows.

D. *inspect* cooling *systems* using ground-source, water-source, solar, and renewable energy technologies.

10. INTERIORS

10.1 The *inspector* shall *inspect*:

A. walls, ceilings, and floors.

B. steps, stairways, and railings.

C. countertops and a *representative number* of *installed* cabinets.

D. a *representative number* of doors and windows.

E. garage vehicle doors and garage vehicle door operators.

F. *installed* ovens, ranges, surface cooking appliances, microwave ovens, dishwashing machines, and food waste grinders by using *normal operating controls* to activate the primary function.

10.2 The *inspector* is NOT required to *inspect*:

A. paint, wallpaper, and other finish treatments.

B. floor coverings.

C. window treatments.

D. coatings on and the hermetic seals between panes of window glass.

E. central vacuum *systems*.

F. *recreational facilities*.

G. *installed* and free-standing kitchen and laundry appliances not listed in Section 10.1.F.

H. appliance thermostats including their calibration, adequacy of heating elements, self cleaning oven cycles, indicator lights, door seals, timers, clocks, timed features, and other specialized features of the appliance.

I. operate, or confirm the operation of every control and feature of an inspected appliance.

11. INSULATION AND VENTILATION

11.1 The *inspector* shall:

  A. *inspect*:

   1. insulation and vapor retarders in unfinished spaces.

   2. ventilation of attics and foundation areas.

   3. kitchen, bathroom, laundry, and similar exhaust *systems*.

   4. clothes dryer exhaust *systems*.

  B. *describe*:

   1. insulation and vapor retarders in unfinished spaces.

   2. absence of insulation in unfinished spaces at conditioned surfaces.

11.2 The *inspector* is NOT required to disturb insulation.

12. FIREPLACES AND FUEL-BURNING APPLIANCES

12.1 The *inspector* shall:

  A. *inspect*:

   1. fuel-burning fireplaces, stoves, and fireplace inserts.

   2. fuel-burning accessories *installed* in fireplaces.

   3. chimneys and vent *systems*.

  B. *describe systems* and *components* listed in 12.1.A.1 and .2.

12.2 The *inspector* is NOT required to:

  A. *inspect*:

   1. interiors of vent *systems*, flues, and chimneys that are not *readily accessible*.

   2. fire screens and doors.

   3. seals and gaskets.

   4. automatic fuel feed devices.

   5. mantles and fireplace surrounds.

   6. combustion air *components* and to determine their adequacy.

   7. heat distribution assists (gravity fed and fan assisted).

   8. fuel-burning fireplaces and appliances located outside the *inspected* structures.

  B. determine draft characteristics.

  C. move fireplace inserts and stoves or firebox contents.

13. GENERAL LIMITATIONS AND EXCLUSIONS

13.1 General limitations

A. The *inspector* is NOT required to perform actions, or to make determinations, or to make recommendations not specifically stated in this Standard.

B. Inspections performed using this Standard:

1. are not *technically exhaustive.*

2. are not required to identify and to report:

a. concealed conditions, latent defects, consequential damages, and

b. cosmetic imperfections that do not significantly affect a *component's* performance of its intended function.

C. This Standard applies to buildings with four or fewer dwelling units and their attached and detached garages and carports.

D. This Standard shall not limit or prevent the *inspector* from meeting state statutes which license professional *home inspection* and home *inspectors.*

E. Redundancy in the description of the requirements, limitations, and exclusions regarding the scope of the *home inspection* is provided for emphasis only.

13.2 General exclusions

A. The *inspector* is NOT required to determine:

1. the condition of *systems* and *components* that are not *readily accessible.*

2. the remaining life expectancy of *systems* and *components.*

3. the strength, adequacy, effectiveness, and efficiency of *systems* and *components.*

4. the causes of conditions and deficiencies.

5. methods, materials, and costs of corrections.

6. future conditions including but not limited to failure of *systems* and *components.*

7. the suitability of the property for specialized uses.

8. compliance of *systems* and *components* with past and present requirements and guidelines (codes, regulations, laws, ordinances, specifications, installation and maintenance instructions, use and care guides, etc.).

9. the market value of the property and its marketability.

10. the advisability of purchasing the property.

11. the presence of plants, animals, and other life forms and substances that may be hazardous or harmful to humans including, but not limited to, wood destroying organisms, molds and mold-like substances.

12. the presence of environmental hazards including, but not limited to, allergens, toxins, carcinogens, electromagnetic radiation, noise, radioactive substances, and contaminants in building materials, soil, water, and air.

13. the effectiveness of *systems installed* and methods used to control or remove suspected hazardous plants, animals, and environmental hazards.

14. operating costs of *systems* and *components*.

15. acoustical properties of *systems* and *components*.

16. soil conditions relating to geotechnical or hydrologic specialties.

17. whether items, materials, conditions and *components* are subject to recall, controversy, litigation, product liability, and other adverse claims and conditions.

B. The *inspector* is NOT required to offer:

1. or to perform acts or services contrary to law or to government regulations.

2. or to perform architectural, *engineering*, contracting, or surveying services or to confirm or to evaluate such services performed by others.

3. or to perform trades or professional services other than *home inspection*.

4. warranties or guarantees.

C. The *inspector* is NOT required to operate:

1. *systems* and *components* that are *shut down* or otherwise inoperable.

2. *systems* and *components* that do not respond to *normal operating controls*.

3. shut-off valves and manual stop valves.

4. *automatic safety controls*.

D. The *inspector* is NOT required to enter:

1. areas that will, in the professional judgment of the *inspector*, likely be dangerous to the *inspector* or to other persons, or to damage the property or its *systems* and *components*.

2. *under-floor crawlspaces* and attics that are not *readily accessible*.

E. The *inspector* is NOT required to inspect:

1. underground items including, but not limited to, underground storage tanks and other underground indications of their presence, whether abandoned or active.

2. items that are not *installed*.

3. *installed decorative* items.

4. items in areas that are not entered in accordance with 13.2.D.

5. detached structures other than garages and carports.

6. common elements and common areas in multi-unit housing, such as condominium properties and cooperative housing.

7. every occurrence of multiple similar *components*.

8. outdoor cooking appliances.

F. The *inspector* is NOT required to:

1. perform procedures or operations that will, in the professional judgment of the *inspector*, likely be dangerous to the *inspector* or to other persons, or to damage the property or its *systems* or *components*.

2. *describe* or report on *systems* and *components* that are not included in this Standard and that were not *inspected*.

3. move personal property, furniture, equipment, plants, soil, snow, ice, and debris.

4. *dismantle systems* and *components*, except as explicitly required by this Standard.

5. reset, reprogram, or otherwise adjust devices, *systems*, and *components* affected by inspection required by this Standard.

6. ignite or extinguish fires, pilot lights, burners, and other open flames that require manual ignition.

7. probe surfaces that would be damaged or where no deterioration is visible or presumed to exist.

14. GLOSSARY OF ITALICIZED TERMS

Automatic Safety Controls   Devices designed and *installed* to protect *systems* and *components* from *unsafe* conditions

Component   A part of a *system*

Decorative   Ornamental; not required for the proper operation of the essential *systems* and *components* of a home

Describe   To identify (in writing) a *system* and *component* by its type or other distinguishing characteristics

Dismantle   To take apart or remove *components*, devices, or pieces of equipment that would not be taken apart or removed by a homeowner in the course of normal maintenance

Engineering   The application of scientific knowledge for the design, control, or use of building structures, equipment, or apparatus

Further Evaluation   Examination and analysis by a qualified professional, tradesman, or service technician beyond that provided by a *home inspection*

Home Inspection   The process by which an *inspector* visually examines the *readily accessible systems* and *components* of a home and *describes* those *systems* and *components* using this Standard

Inspect   The process of examining *readily accessible systems* and *components* by (1) applying this Standard, and (2) operating *normal operating controls*, and (3) opening *readily openable access panels*

Inspector   A person hired to examine *systems* and *components* of a building using this Standard

Installed   Attached such that removal requires tools

Normal Operating Controls   Devices such as thermostats, switches, and valves intended to be operated by the homeowner

Readily Accessible   Available for visual *inspection* without requiring moving of personal property, *dismantling*, destructive measures, or actions that will likely involve risk to persons or property

Readily Openable Access Panel   A panel provided for homeowner inspection and maintenance that is *readily accessible*, within normal reach, can be opened by one person, and is not sealed in place

Recreational Facilities   Spas, saunas, steam baths, swimming pools, exercise, entertainment, athletic, playground and other similar equipment, and associated accessories

Representative Number   One *component* per room for multiple similar interior *components* such as windows and electric receptacles; one component on each side of the building for multiple similar exterior components

Roof Drainage Systems   *Components* used to carry water off a roof and away from a building

Shut Down   A state in which a *system* or *component* cannot be operated by *normal operating controls*

Structural Component   A *component* that supports non-variable forces or weights (dead loads) and variable forces or weights (live loads)

System   A combination of interacting or interdependent *components*, assembled to carry out one or more functions

Technically Exhaustive   An investigation that involves *dismantling*, the extensive use of advanced techniques, measurements, instruments, testing, calculations, or other means

Under-floor Crawlspace   The area within the confines of the foundation and between the ground and the underside of the floor

Unsafe   A condition in a *readily accessible, installed system* or *component* that is judged by the *inspector* to be a significant risk of serious bodily injury during normal, day-to-day use; the risk may be due to damage, deterioration, improper installation, or a change in accepted residential construction practices

Wall Covering A protective or insulating layer fixed to the outside of a building such as: aluminum, brick, EIFS, stone, stucco, vinyl, and wood

Wiring Method Identification of electrical conductors or wires by their general type, such as non-metallic sheathed cable, armored cable, and knob and tube, etc.

Index

Other Books by Bruce A. Barker

Bruce A. Barker was Chair of the ASHI Standards Committee when the Committee revised the home inspection standard. He remains actively involved in ASHI standards development.

Other books by this author include:

The NHIE Home Inspection Manual

Everybody's Building Code (various editions)

Black & Decker Codes for Homeowners (various editions)

Black & Decker Deck Codes and Standards

Books edited by this author include:

Black & Decker Complete Guide to Wiring

Black & Decker Advanced Home Wiring

Black & Decker Complete Guide to Plumbing